T0325085

PRIME NUMBERS AND THE RIEMANN HYPOTHESIS

Prime numbers are beautiful, mysterious, and beguiling mathematical objects. The mathematician Bernhard Riemann made a celebrated conjecture about primes in 1859, the so-called Riemann Hypothesis, which remains to be one of the most important unsolved problems in mathematics. Through the deep insights of the authors, this book introduces primes and explains the Riemann Hypothesis.

Students with minimal mathematical background and scholars alike will enjoy this comprehensive discussion of primes. The first part of the book will inspire the curiosity of a general reader with an accessible explanation of the key ideas. The exposition of these ideas is generously illuminated by computational graphics that exhibit the key concepts and phenomena in enticing detail. Readers with more mathematical experience will then go deeper into the structure of primes and see how the Riemann Hypothesis relates to Fourier analysis using the vocabulary of spectra. Readers with a strong mathematical background will be able to connect these ideas to historical formulations of the Riemann Hypothesis.

Barry Mazur is the Gerhard Gade University Professor at Harvard University. He is the author of *Imagining Numbers (particularly the square root of minus fifteen)* and coeditor, with Apostolos Doxiadis, of *Circles Disturbed: The Interplay of Mathematics and Narrative.*

William Stein is Professor of Mathematics at the University of Washington. Author of *Elementary Number Theory: Primes, Congruences, and Secrets: A Computational Approach,* he is also the founder of the Sage mathematical software project.

Prime Numbers and the Riemann Hypothesis

BARRY MAZUR

Harvard University, Cambridge, MA, USA

WILLIAM STEIN

University of Washington, Seattle, WA, USA

CAMBRIDGE
UNIVERSITY PRESS

32 Avenue of the Americas, New York NY 10013-2473, USA

Cambridge University Press is part of the University of Cambridge.

It furthers the University's mission by disseminating knowledge in the pursuit of education, learning and research at the highest international levels of excellence.

www.cambridge.org
Information on this title: www.cambridge.org/9781107101920

© Barry Mazur and William Stein 2016

First published 2016

A catalogue record for this publication is available from the British Library

Library of Congress Cataloguing in Publication data
Mazur, Barry.
Prime numbers and the Riemann hypothesis / Barry Mazur, Harvard University, Cambridge, MA, USA, William Stein, University of Washington, Seattle, WA, USA.
Includes bibliographical references and index.
ISBN 978-1-107-10192-0 (hardback : alk. paper) – ISBN 978-1-107-49943-0 (pbk. : alk. paper)
1. Riemann hypothesis.   2. Numbers, Prime.   I. Stein, William A., 1974–   II. Title.
QA246.M49   2015
512.7'3 – dc23   2015018981

ISBN 978-1-107-10192-0 Hardback

Contents

Preface

The Riemann Hypothesis is one of the great unsolved problems of mathematics, and the reward of $1,000,000 of *Clay Mathematics Institute* prize money awaits the person who solves it. But – with or without money – its resolution is crucial for our understanding of the nature of numbers.

There are several full-length books recently published, written for a general audience, that have the Riemann Hypothesis as their main topic. A reader of these books will get a fairly rich picture of the personalities engaged in the pursuit, and of related mathematical and historical issues.[1]

This is *not* the mission of the book that you now hold in your hands. We aim – instead – to explain, in as direct a manner as possible and with the least mathematical background required, what this problem is all about and why it is so important. For even before anyone proves this *hypothesis* to be true (or false!), just getting familiar with it, and with some of the ideas behind it, is exciting. Moreover, this hypothesis is of crucial importance in a wide range of mathematical fields; for example, it is a confidence-booster for computational mathematics: even if the Riemann Hypothesis is never proved, assuming its truth (and that of closely related hypotheses) gives us an excellent sense of how long certain computer programs will take to run, which, in some cases, gives us the assurance we need to initiate a computation that might take weeks or even months to complete.

Here is how the Princeton mathematician Peter Sarnak describes the broad impact the Riemann Hypothesis has had[2]:

"The Riemann hypothesis is the central problem and it implies many, many things. One thing that makes it rather unusual in mathematics today is that there must be over five hundred papers – somebody should

---

[1] See, e.g., *The Music of the Primes* by Marcus du Sautoy (2003) and *Prime Obsession: Bernhard Riemann and the Greatest Unsolved Problem in Mathematics* by John Derbyshire (2003).

[2] See page 222 of *The Riemann hypothesis: the greatest unsolved problem in mathematics* by Karl Sabbagh (2002).

Figure 0.1. Peter Sarnak. Photo by William Stein

go and count – which start 'Assume the Riemann hypothesis[3],' and the conclusion is fantastic. And those [conclusions] would then become theorems . . . With this one solution you would have proven five hundred theorems or more at once."

So, what *is* the Riemann Hypothesis? Below is a *first description* of what it is about. The task of our book is to develop the following boxed paragraph into a fuller explanation and to convince you of the importance and beauty of the mathematics it represents. We will be offering, throughout our book, a number of different – but equivalent – ways of precisely formulating this hypothesis (we display these in boxes). When we say that two mathematical statements are "equivalent," we mean that, given the present state of mathematical knowledge, we can prove that if either one of those statements is true, then the other is true. The endnotes will guide the reader to the relevant mathematical literature.

What sort of Hypothesis is the Riemann Hypothesis?

Consider the seemingly innocuous series of questions:

- How many prime numbers (2, 3, 5, 7, 11, 13, 17, . . .) are there less than 100?
- How many less than 10,000?
- How many less than 1,000,000?

More generally, how many primes are there less than any given number X?

---

[3] Technically, a generalized version of the Riemann hypothesis (see Chapter 38 below).

Riemann proposed, a century and half ago, a strikingly simple-to-describe "very good approximation" to the number of primes less than or equal to a given number X. We now see that if we could prove this *Hypothesis of Riemann* we would have the key to a wealth of powerful mathematics. Mathematicians are eager to find that key.

Figure 0.2. Raoul Bott. Courtesy of the Department of Mathematics, Harvard University

The mathematician Raoul Bott – in giving advice to a student – once said that whenever one reads a mathematics book or article, or goes to a math lecture, one should aim to come home with something very specific (it can be small, but should be *specific*) that has application to a wider class of mathematical problems than was the focus of the text or lecture. If we were to suggest some possible *specific* items to come home with, after reading our book, three key phrases – **prime numbers**, **square-root accurate**, and **spectrum** – would head the list. As for words of encouragement to think hard about the first of these, i.e., prime numbers, we can do no better than to quote a paragraph of Don Zagier's classic 12-page exposition, *The First 50 Million Prime Numbers*:

> "There are two facts about the distribution of prime numbers of which I hope to convince you so overwhelmingly that they will be permanently engraved in your hearts. The first is that, [they are] the most arbitrary and ornery objects studied by mathematicians: they grow like weeds among the natural numbers, seeming to obey no other law than that of chance, and nobody can predict where the next one will sprout. The second fact is even more astonishing, for it states just the opposite: that the

Figure 0.3. Don Zagier. Photo by William Stein

prime numbers exhibit stunning regularity, that there are laws governing their behavior, and that they obey these laws with almost military
precision."

Mathematics is flourishing. Each year sees new exciting initiatives that
extend and sharpen the applications of our subject, new directions for deep
exploration – and finer understanding – of classical as well as very contemporary mathematical domains. We are aided in such explorations by the development of more and more powerful tools. We see resolutions of centrally important questions. And through all of this, we are treated to surprises and dramatic
changes of viewpoint; in short: marvels.

And what an array of wonderful techniques allow mathematicians to do
their work: framing *definitions;* producing *constructions;* formulating *analogies
relating disparate concepts, and disparate mathematical fields;* posing *conjectures,* that cleanly shape a possible way forward; and, the keystone: providing
unassailable *proofs* of what is asserted, the idea of doing such a thing being
itself one of the great glories of mathematics.

Number theory has its share of this bounty. Along with all these modes of
theoretical work, number theory also offers the pure joy of numerical experimentation, which – when it is going well – allows you to witness the intricacy of
numbers and profound inter-relations that cry out for explanation. It is striking how little you actually have to know in order to appreciate the revelations
offered by numerical exploration.

Our book is meant to be an introduction to these pleasures. We take an
experimental view of the fundamental ideas of the subject buttressed by
numerical computations, often displayed as graphs. As a result, our book is

profusely illustrated, containing 131 figures, diagrams, and pictures that accompany the text.[4]

There are few mathematical equations in Part I. This first portion of our book is intended for readers who are generally interested in, or curious about, mathematical ideas, but who may not have studied any advanced topics. Part I is devoted to conveying the essence of the Riemann Hypothesis and explaining why it is so intensely pursued. It requires a minimum of mathematical knowledge, and does not, for example, use calculus, although it would be helpful to know – or to learn on the run – the meaning of the concept of *function*. Given its mission, Part I is meant to be complete, in that it has a beginning, middle, and end. We hope that our readers who only read Part I will have enjoyed the excitement of this important piece of mathematics.

Part II is for readers who have taken at least one class in calculus, possibly a long time ago. It is meant as a general preparation for the type of Fourier analysis that will occur in the later parts. The notion of spectrum is key.

Part III is for readers who wish to see, more vividly, the link between the placement of prime numbers and (what we call there) the *Riemann spectrum*.

Part IV requires some familiarity with complex analytic functions, and returns to Riemann's original viewpoint. In particular it relates the "Riemann spectrum" that we discuss in Part III to the *nontrivial zeroes of the Riemann zeta function*. We also provide a brief sketch of the more standard route taken by published expositions of the Riemann Hypothesis.

The end-notes are meant to link the text to references, but also to provide more technical commentary with an increasing dependence on mathematical background in the later chapters. References to the end notes will be in brackets.

We wrote our book over the past decade, but devoted only one week to it each year (a week in August). At the end of our work-week for the book, each year, we put our draft (mistakes and all) on line to get response from readers.[5] We therefore accumulated much important feedback, corrections, and requests from readers, especially J. S. Markovitch who very carefully proofread the final draft.[6] We thank them infinitely.

---

[4] We created the figures using the free SageMath software (see `http://www.sagemath.org`). Complete source code is available, which can be used to recreate every diagram in this book (see `http://wstein.org/rh`). More adventurous readers can try to experiment with the parameters for the ranges of data illustrated, so as to get an even more vivid sense of how the numbers "behave." We hope that readers become inspired to carry out numerical experimentation, which is becoming easier as mathematical software advances.

[5] See `http://library.fora.tv/2014/04/25/Riemann_Hypothesis_The_Million_Dollar_Challenge` which is a lecture – and Q & A – about the composition of this book.

[6] Including Dan Asimov, Bret Benesh, Keren Binyaminov, Harald Bögeholz, Louis-Philippe Chiasson, Keith Conrad, Karl-Dieter Crisman, Nicola Dunn, Thomas Egense, Bill Gosper, Andrew Granville, Shaun Griffith, Michael J. Gruber, Robert Harron, William R. Hearst III, David Jao, Fredrik Johansson, Jim Markovitch, David Mumford, James Propp, Andrew Solomon, Dennis Stein, and Chris Swenson.

PART I

The Riemann Hypothesis

1 Thoughts About Numbers: Ancient, Medieval, and Modern

If we are to believe the ancient Greek philosopher Aristotle, the early Pythagoreans thought that the principles governing Number are "the principles of all things," the concept of Number being more basic than *earth, air, fire, or water*, which were according to ancient tradition the four building blocks of matter. To think about Number is to get close to the architecture of "what is."

So, how far along are we in our thoughts about numbers?

Figure 1.1. René Descartes (1596–1650) © RMN-Grand Palais / Art Resource, NY

The French philosopher and mathematician René Descartes, almost four centuries ago, expressed the hope that there soon would be "almost nothing more to discover in geometry." Contemporary physicists dream of a final

Figure 1.2. Jean de Bosschere, "Don Quixote and his Dulcinea del Toboso," from The History of Don Quixote De La Mancha, by Miguel De Cervantes. Trans. Thomas Shelton. Constable and Company, New York, 1922

theory.[1] But despite its venerability and its great power and beauty, the pure mathematics of numbers may still be in the infancy of its development, with depths to be explored as endless as the human soul, and *never* a final theory.

Numbers are obstreperous things. Don Quixote encountered this when he requested that the "bachelor" compose a poem to his lady Dulcinea del Toboso, the first letters of each line spelling out her name. The "bachelor" found[2]

> "a great difficulty in their composition because the number of letters in her name was 17, and if he made four Castilian stanzas of four octo-syllabic lines each, there would be one letter too many, and if he made the stanzas of five octosyllabic lines each, the ones called *décimas* or *redondillas,* there would be three letters too few..."

"It must fit in, however you do it," pleaded Quixote, not willing to grant the imperviousness of the number 17 to division.

Seventeen is indeed a prime number: there is no way of factoring it as the product of smaller numbers, and this accounts – people tell us – for its occurrence in some phenomena of nature, as when the seventeen-year cicadas all emerged to celebrate a "reunion" of some sort in our fields and valleys.

Prime numbers, despite their *primary* position in our modern understanding of numbers, were not specifically doted over in the ancient literature before Euclid, at least not in the literature that has been preserved. Primes are mentioned as a class of numbers in the writings of Philolaus (a predecessor of Plato);

---

[1] See Weinberg's book *Dreams of a Final Theory: The Search for the Fundamental Laws of Nature*, by Steven Weinberg (New York: Pantheon Books, 1992).

[2] See Chapter IV of the Second Part of the *Ingenious Gentleman Don Quixote of La Mancha.*

Figure 1.3. Cicadas emerge every 17 years. Photo by Bob Peterson

they are not mentioned specifically in the Platonic dialogues, which is surprising given the intense interest Plato had in mathematical developments; and they make an occasional appearance in the writings of Aristotle, which is not surprising, given Aristotle's emphasis on the distinction between the *composite* and the *incomposite*. "The incomposite is prior to the composite," writes Aristotle in Book 13 of the Metaphysics.

Prime numbers do occur, in earnest, in Euclid's *Elements*!

There is an extraordinary wealth of established truths about whole numbers; these truths provoke sheer awe for the beautiful complexity of prime numbers. But each of the important new discoveries we make gives rise to a further richness of questions, educated guesses, heuristics, expectations, and unsolved problems.

2 | What are Prime Numbers?

Primes as atoms. To begin from the beginning, think of the operation of multiplication as a bond that ties numbers together: the equation $2 \times 3 = 6$ invites us to imagine the number 6 as (a molecule, if you wish) built out of its smaller constituents 2 and 3. Reversing the procedure, if we start with a whole number, say 6 again, we may try to factor it (that is, express it as a product of smaller whole numbers) and, of course, we would eventually, if not immediately, come up with $6 = 2 \times 3$ and discover that 2 and 3 factor no further; the numbers 2 and 3, then, are the indecomposable entities (atoms, if you wish) that comprise our number.

Figure 2.1. The number $6 = 2 \times 3$

By definition, a **prime number** (colloquially, *a prime*) is a whole number, bigger than 1, that cannot be factored into a product of two smaller whole numbers. So, 2 and 3 are the first two prime numbers. The next number along the line, 4, is not prime, for $4 = 2 \times 2$; the number after that, 5, is. Primes are, multiplicatively speaking, the building blocks from which all numbers can be made. A fundamental theorem of arithmetic tells us that any number (bigger than 1) can be factored as a product of primes, and the factorization is *unique* except for rearranging the order of the primes.

For example, if you try to factor 12 as a product of two smaller numbers – ignoring the order of the factors – there are two ways to begin to do this:

$$12 = 2 \times 6 \quad \text{and} \quad 12 = 3 \times 4$$

But neither of these ways is a full factorization of 12, for both 6 and 4 are not prime, so can be themselves factored, and in each case after changing the ordering of the factors we arrive at:

$$12 = 2 \times 2 \times 3.$$

If you try to factor the number 300, there are many ways to begin:

$$300 = 30 \times 10 \quad \text{or} \quad 300 = 6 \times 50$$

and there are various other starting possibilities. But if you continue the factorization ("climbing down" any one of the possible "factoring trees") to the bottom, where every factor is a prime number as in Figure 2.2, you always end up with the same collection of prime numbers[1]:

$$300 = 2^2 \times 3 \times 5^2.$$

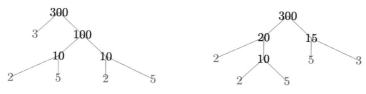

Figure 2.2. Factor trees that illustrate the factorization of 300 as a product of primes.

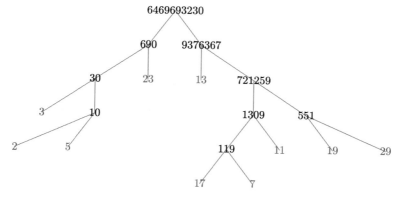

Figure 2.3. Factorization tree for the product of the primes up to 29.

The Riemann Hypothesis probes the question: how intimately can we know prime numbers, those *atoms* of multiplication? Prime numbers are an

---

[1] See Section 1.1 of Stein's *Elementary Number Theory: Primes, Congruences, and Secrets* (2008) at http://wstein.org/ent/ for a proof of the "fundamental theorem of arithmetic", which asserts that every positive whole number factors uniquely as a product of primes.

important part of our daily lives. For example, often when we visit a website and purchase something online, prime numbers having hundreds of decimal digits are used to keep our bank transactions private. This ubiquitous use to which giant primes are put depends upon a very simple principle: it is much easier to multiply numbers together than to factor them. If you had to factor, say, the number 391 you might scratch your head for a few minutes before discovering that 391 is 17×23. But if you had to multiply 17 by 23 you would do it straightaway. Offer two primes, say, P and Q each with a few hundred digits, to your computing machine and ask it to multiply them together: you will get their product $N = P \times Q$ with its hundreds of digits in about a microsecond. But present that number N to any current desktop computer, and ask it to factor N, and the computer will (almost certainly) fail to do the task. See [1] and [2].

The safety of much encryption depends upon this "guaranteed" failure![2]

If we were latter-day number-phenomenologists we might revel in the discovery and proof that

$$p = 2^{43,112,609} - 1 = 3164702693 \ldots \ldots \text{(millions of digits)} \ldots \ldots 6697152511$$

is a prime number, this number having 12,978,189 digits! This prime, which was discovered on August 23, 2008 by the GIMPS project,[3] is the first prime ever found with more than ten million digits, though it is not the largest prime currently known.

Now $2^{43,112,609} - 1$ is quite a hefty number! Suppose someone came up to you saying "surely $p = 2^{43,112,609} - 1$ is the largest prime number!" (which it is not). How might you convince that person that he or she is wrong without explicitly exhibiting a larger prime? [3]

Here is a neat – and, we hope, convincing – strategy to show there are prime numbers larger than $p = 2^{43,112,609} - 1$. Imagine forming the following humongous number: let M be the product of all prime numbers up to and including $p = 2^{43,11,2609} - 1$. Now go one further than M by taking the next number $N = M + 1$.

OK, even though this number N is wildly large, it is either a prime number itself – which would mean that there would indeed be a prime number larger than $p = 2^{43,112,609} - 1$, namely N; or in any event it is surely divisible by some prime number, call it P.

Here, now, is a way of seeing that this P is bigger than p. Since every prime number smaller than or equal to p divides M, these prime numbers cannot divide $N = M + 1$ (since they divide M evenly, if you tried to divide $N = M + 1$ by any of them you would get a remainder of 1). So, since P does divide N it must not be any of the smaller prime numbers: P is therefore a prime number bigger than $p = 2^{43,112,609} - 1$.

---

[2] Nobody has ever published a *proof* that there is no fast way to factor integers. This is an article of "faith" among some cryptographers.

[3] The GIMPS project website is http://www.mersenne.org/.

This strategy, by the way, is not very new: it is, in fact, well over two thousand years old, since it already occurred in Euclid's *Elements*. The Greeks did know that there are infinitely many prime numbers and they showed it via the same method as we showed that our $p = 2^{43,112,609} - 1$ is not the largest prime number.

Here is the argument again, given very succinctly: Given primes $p_1, \ldots, p_m$, let $n = p_1 p_2 \cdots p_m + 1$. Then n is divisible by some prime not equal to any p_i, so there are more than m primes.

You can think of this strategy as a simple game that you can play. Start with the bag of prime numbers that contains just the two primes 2 and 3. Now each "move" of the game consists of multiplying together all the primes you have in your bag to get a number M, then adding 1 to M to get the larger number $N = M + 1$, then factoring N into prime number factors, and then including all those new prime numbers in your bag. Euclid's proof gives us that we will – with each move of this game – be finding more prime numbers: the contents of the bag will increase. After, say, a million moves our bag will be guaranteed to contain more than a million prime numbers.

For example, starting the game with your bag containing only one prime number 2, here is how your bag grows with successive moves of the game:

$\{2\}$

$\{2, 3\}$

$\{2, 3, 7\}$

$\{2, 3, 7, 43\}$

$\{2, 3, 7, 43, 13, 139\}$

$\{2, 3, 7, 43, 13, 139, 3263443\}$

$\{2, 3, 7, 43, 13, 139, 3263443, 547, 607, 1033, 31051\}$

$\{2, 3, 7, 43, 13, 139, 3263443, 547, 607, 1033, 31051, 29881, 67003,$
$\qquad 9119521, 6212157481\}$

etc.[4]

Though there are infinitely many primes, explicitly finding large primes is a major challenge. In the 1990s, the Electronic Frontier Foundation `http://www` `.eff.org/awards/coop` offered a \$100,000 cash reward to the first group to find a prime with at least 10,000,000 decimal digits (the group that found the record prime p above won this prize[5]), and offers another \$150,000 cash prize to the first group to find a prime with at least 100,000,000 decimal digits.

The number $p = 2^{43,112,609} - 1$ was for a time the largest prime known, where by "know" we mean that we know it so explicitly that we can *compute* things

---

[4] The sequence of prime numbers we find by this procedure is discussed in more detail with references in the Online Encyclopedia of Integer Sequences `http://oeis.org/` `A126263`.

[5] See `http://www.eff.org/press/archives/2009/10/14-0`. Also the 46th Mersenne prime was declared by *Time Magazine* to be one of the top 50 best "inventions" of 2008: `http://www.time.com/time/specials/packages/article/0,` `28804,1852747_1854195_1854157,00.html`.

about it. For example, the last two digits of p are both 1 and the sum of the digits of p is 58,416,637. Of course p is not the largest prime number since there are infinitely many primes, e.g., the next prime q after p is a prime. But there is no known way to efficiently compute anything interesting about q. For example, what is the last digit of q in its decimal expansion?

3 | "Named" Prime Numbers

Prime numbers come in all sorts of shapes, some more convenient to deal with than others. For example, the number we have been talking about,

$$p = 2^{43,112,609} - 1,$$

is given to us, by its very notation, in a striking form; i.e., *one less than a power of 2*. It is no accident that the largest "currently known" prime number has such a form. This is because there are special techniques we can draw on to show primality of a number, if it is one less than a power of 2 and – of course – if it also happens to be prime. The primes of that form have a name, *Mersenne Primes*, as do the primes that are *one more than a power of 2*, those being called *Fermat Primes*. [4]

Here are two exercises that you might try to do, if this is your first encounter with primes that differ from a power of 2 by 1:

1. Show that if a number of the form $M = 2^n - 1$ is prime, then the exponent n is also prime. [Hint: This is equivalent to proving that if n is composite, then $2^n - 1$ is also composite.] For example: $2^2 - 1 = 3$, $2^3 - 1 = 7$ are primes, but $2^4 - 1 = 15$ is not. So *Mersenne primes* are numbers that are
   - of the form $2^{\text{prime number}} - 1$, and
   - are themselves prime numbers.
2. Show that if a number of the form $F = 2^n + 1$ is prime, then the exponent n is a power of two. For example: $2^2 + 1 = 5$ is prime, but $2^3 + 1 = 9$ is not. So *Fermat primes* are numbers that are
   - of the form $2^{\text{power of two}} + 1$, and
   - are themselves prime numbers.

Not all numbers of the form $2^{\text{prime number}} - 1$ or of the form $2^{\text{power of two}} + 1$ are prime. We currently know only finitely many primes of either of these forms. How we have come to know what we know is an interesting tale. See, for example, `http://www.mersenne.org/`.

4 Sieves

Eratosthenes, the mathematician from Cyrene (and later, librarian at Alexandria) explained how to *sift* the prime numbers from the series of all numbers: in the sequence of numbers,

2 3 4 5 6 7 8 9 10 11 12 13 14 15 16 17 18 19 20 21 22 23 24 25 26,

for example, start by circling the 2 and crossing out all the other multiples of 2. Next, go back to the beginning of our sequence of numbers and circle the first number that is neither circled nor crossed out (that would be, of course, the 3), then cross out all the other multiples of 3. This gives the pattern: go back again to the beginning of our sequence of numbers and circle the first number that is neither circled nor crossed out; then cross out all of its other multiples. Repeat this pattern until all the numbers in our sequence are either circled, or crossed out, the circled ones being the primes.

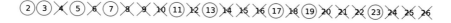

In Figures 4.1–4.4 we use the primes 2, 3, 5, and finally 7 to sieve out the primes up to 100, where instead of crossing out multiples we grey them out, and instead of circling primes we color their box red.

Since all the even numbers greater than two are eliminated as being composite numbers and not primes they appear as gray in Figure 4.1, but none of the odd numbers are eliminated so they still appear in white boxes.

Looking at Figure 4.3, we see that for all but three numbers (49, 77, and 91) up to 100 we have (after sieving by 2,3, and 5) determined which are primes and which composite.

	2	3	4	5	6	7	8	9	10
11	12	13	14	15	16	17	18	19	20
21	22	23	24	25	26	27	28	29	30
31	32	33	34	35	36	37	38	39	40
41	42	43	44	45	46	47	48	49	50
51	52	53	54	55	56	57	58	59	60
61	62	63	64	65	66	67	68	69	70
71	72	73	74	75	76	77	78	79	80
81	82	83	84	85	86	87	88	89	90
91	92	93	94	95	96	97	98	99	100

Figure 4.1. Using the prime 2 to sieve for primes up to 100

	2	3		5		7		9	
11		13		15		17		19	
21		23		25		27		29	
31		33		35		37		39	
41		43		45		47		49	
51		53		55		57		59	
61		63		65		67		69	
71		73		75		77		79	
81		83		85		87		89	
91		93		95		97		99	

Figure 4.2. Using the primes 2 and 3 to sieve for primes up to 100

	2	3		5		7			
11		13				17		19	
		23		25				29	
31				35		37			
41		43				47		49	
		53		55				59	
61				65		67			
71		73				77		79	
		83		85				89	
91				95		97			

Figure 4.3. Using the primes 2, 3, and 5 to sieve for primes up to 100

	2	3		5		7			
11		13				17		19	
		23						29	
31						37			
41		43				47		49	
		53						59	
61						67			
71		73				77		79	
		83						89	
91						97			

Figure 4.4. Using the primes 2, 3, 5, and 7 to sieve for primes up to 100

Finally, we see in Figure 4.4 that sieving by 2, 3, 5, and 7 determines all primes up to 100. See [5] for more about explicitly enumerating primes using a computer.

5 | Questions About Primes that any Person Might Ask

We become quickly stymied when we ask quite elementary questions about the spacing of the infinite series of prime numbers.

For example, *are there infinitely many pairs of primes whose difference is* 2*?* The sequence of primes seems to be rich in such pairs

$$5 - 3 = 2, \quad 7 - 5 = 2, \quad 13 - 11 = 2, \quad 19 - 17 = 2,$$

and we know that there are loads more such pairs[1] but the answer to our question, *are there infinitely many?*, is not known. The conjecture that there are infinitely many such pairs of primes ("twin primes" as they are called) is known as the *Twin Primes Conjecture. Are there infinitely many pairs of primes whose difference is* 4, 6*?* Answer: equally unknown. Nevertheless there is very exciting recent work in this direction, specifically, Yitang Zhang proved that there are infinitely many pairs of primes that differ by no more than 7×10^7. For a brief account of Zhang's work, see the Wikipedia entry http://en.wikipedia.org/wiki/Yitang_Zhang. Many exciting results have followed Zhang's breakthrough; we know now, thanks to results[2] of James Maynard and others, that there are infinitely many pairs of primes that differ by no more than 246.

---

[1] For example, according to http://oeis.org/A007508 there are $10,304,185,697,298$ such pairs less than $10,000,000,000,000,000$.
[2] See https://www.simonsfoundation.org/quanta/20131119-together-and-alone-closing-the-prime-gap/ and for further work http://michaelnielsen.org/polymath1/index.php?title=Bounded_gaps_between_primes.

Is every even number greater than 2 a sum of two primes? Answer: unknown. *Are there infinitely many primes which are 1 more than a perfect square?* Answer: unknown.

Figure 5.1. Yitang Zhang. Portrait courtesy of the University of New Hampshire

Remember the Mersenne prime $p = 2^{43,112,609} - 1$ from Chapter 3 and how we proved – by pure thought – that there must be a prime P larger than p? Suppose, though, someone asked us whether there was a *Mersenne Prime* larger than this p: that is, *is there a prime number of the form*

$$2^{\text{some prime number}} - 1$$

bigger than $p = 2^{43,112,609} - 1$? Answer: For many years we did not know; however, in 2013 Curtis Cooper discovered the even bigger Mersenne prime $2^{57,885,161} - 1$, with a whopping 17,425,170 digits! Again we can ask if there is a Mersenne prime larger than Cooper's. Answer: we do not know. It is possible that there are infinitely many Mersenne primes but we're far from being able to answer such questions.

Is there some neat formula giving the next prime? More specifically, *if I give you a number N, say N = one million, and ask you for the first number after N that is prime, is there a method that answers that question without, in some form or other, running through each of the successive odd numbers after N rejecting the nonprimes until the first prime is encountered?* Answer: unknown.

One can think of many ways of "getting at" some understanding of the placement of prime numbers among all numbers. Up to this point we have been mainly just counting them, trying to answer the question "how many primes

Figure 5.2. Marin Mersenne (1588–1648) © HIP / Art Resource, NY

are there up to X?" and we have begun to get some feel for the numbers behind this question, and especially for the current "best guesses" about estimates.

What is wonderful about this subject is that people attracted to it cannot resist asking questions that lead to interesting, and sometimes surprising numerical experiments. Moreover, given our current state of knowledge, many of the questions that come to mind are still unapproachable: we don't yet know enough about numbers to answer them. But *asking interesting questions* about the mathematics that you are studying is a high art, and is probably a necessary skill to acquire, in order to get the most enjoyment – and understanding – from mathematics. So, we offer this challenge to you:

Come up with with your own question about primes that

- is interesting to you,
- is not a question whose answer is known to you,
- is not a question that you've seen before; or at least not exactly,
- is a question about which you can begin to make numerical investigations.

If you are having trouble coming up with a question, read on for more examples that provide further motivation.

Further Questions About Primes

6

In celebration of Yitang Zhang's recent result, let us consider more of the numerics regarding *gaps* between one prime and the next, rather than the tally of all primes. Of course, it is no fun at all to try to guess how many pairs of primes p, q there are with gap $q - p$ equal to a fixed odd number, since the difference of two odd numbers is even, as in Chapter 5. The fun, though, begins in earnest if you ask for pairs of primes with difference equal to 2 (these being called *twin primes*) for it has long been guessed that there are infinitely many such pairs of primes, but no one has been able to prove this yet.

As of 2014, the largest known twin primes are

$$3756801695685 \cdot 2^{666669} \pm 1.$$

These enormous primes, which were found in 2011, have 200,700 digits each.[1]

Similarly, it is interesting to consider primes p and q with difference 4, or 8, or – in fact – any even number $2k$. That is, people have guessed that there are infinitely many pairs of primes with difference 4, with difference 6, etc. but none of these guesses have yet been proved.

So, define

$$\mathrm{Gap}_k(X)$$

to be the number of pairs of *consecutive* primes (p, q) with $q < X$ that have "gap k" (i.e., such that their difference $q - p$ is k). Here p is a prime, $q > p$ is a prime, and there are no primes between p and q. For example, $\mathrm{Gap}_2(10) = 2$, since the pairs $(3, 5)$ and $(5, 7)$ are the pairs less than 10 with gap 2, and $\mathrm{Gap}_4(10) = 0$ because despite 3 and 7 being separated by 4, they are not consecutive primes.

---

[1] See http://primes.utm.edu/largest.html#twin for the top ten largest known twin primes.

See Table 6.1 for various values of $\text{Gap}_k(X)$ and Figure 6.1 for the distribution of prime gaps for $X = 10^7$.

Table 6.1. Values of $\text{Gap}_k(X)$

X	$\text{Gap}_2(X)$	$\text{Gap}_4(X)$	$\text{Gap}_6(X)$	$\text{Gap}_8(X)$	$\text{Gap}_{100}(X)$	$\text{Gap}_{246}(X)$
10	2	0	0	0	0	0
10^2	8	7	7	1	0	0
10^3	35	40	44	15	0	0
10^4	205	202	299	101	0	0
10^5	1224	1215	1940	773	0	0
10^6	8169	8143	13549	5569	2	0
10^7	58980	58621	99987	42352	36	0
10^8	440312	440257	768752	334180	878	0

The recent results of Zhang as sharpened by Maynard (and others) we mentioned above tell us that for at least one even number k among the even numbers $k \leq 246$, $\text{Gap}_k(X)$ goes to infinity as X goes to infinity. One expects that this happens for *all* even numbers k. We expect this as well, of course, for $\text{Gap}_{246}(X)$ despite what might be misconstrued as discouragement by the above data.

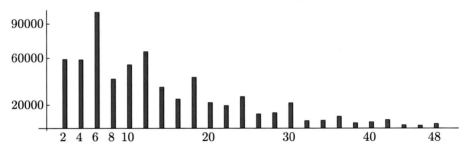

Figure 6.1. Frequency histogram showing the distribution of prime gaps of size ≤ 50 for all primes up to 10^7. Six is the most popular gap in this data.

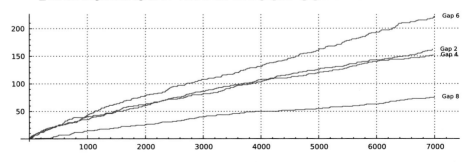

Figure 6.2. Plots of $\text{Gap}_k(X)$ for $k = 2, 4, 6, 8$. Which wins?

Here is yet another question that deals with the spacing of prime numbers that we do not know the answer to:

Racing Gap 2, Gap 4, Gap 6, and Gap 8 against each other:

Challenge: As X tends to infinity which of $\text{Gap}_2(X)$, $\text{Gap}_4(X)$, $\text{Gap}_6(X)$, or $\text{Gap}_8(X)$ do you think will grow faster? How much would you bet on the truth of your guess? [6]

Here is a curious question that you can easily begin to check out for small numbers. We know, of course, that the *even* numbers and the *odd* numbers are nicely and simply distributed: after every odd number comes an even number, after every even, an odd. There are an equal number of positive odd numbers and positive even numbers less than any given odd number, and there may be nothing else of interest to say about the matter. Things change considerably, though, if we focus our concentration on *multiplicatively even* numbers and *multiplicatively odd* numbers.

A **multiplicatively even** number is one that can be expressed as a product of *an even number of* primes; and a **multiplicatively odd** number is one that can be expressed as a product of *an odd number of* primes. So, any prime is multiplicatively odd, the number $4 = 2 \cdot 2$ is multiplicatively even, and so is $6 = 2 \cdot 3$, $9 = 3 \cdot 3$, and $10 = 2 \cdot 5$; but $12 = 2 \cdot 2 \cdot 3$ is multiplicatively odd. Below we list the numbers up to 25, and underline and bold the multiplicatively odd numbers.

1 **2 3** 4 **5** 6 **7 8** 9 10 **11 12 13** 14 15 16 **17 18 19 20** 21 22 **23** 24 25

Table 6.2 gives some data:

Table 6.2. Count of multiplicatively odd and even positive numbers $\leq X$

X	1	2	3	4	5	6	7	8	9	10	11	12	13	14	15	16
m. odd	0	1	2	2	3	3	4	5	5	5	6	7	8	8	8	8
m. even	1	1	1	2	2	3	3	3	4	5	5	5	5	6	7	8

Now looking at this data, a natural, and simple, question to ask about the concept of multiplicative *oddness* and *evenness* is:

Is there some $X \geq 2$ for which there are more multiplicatively even numbers less than or equal to X than multiplicatively odd ones?

Each plot in Figure 6.3 gives the number of multiplicatively even numbers between 2 and X minus the number of multiplicatively odd numbers between 2 and X, for X equal to 10, 100, 1000, 10000, 100000, and 1000000. The above question asks whether these graphs would, for sufficiently large X, ever cross the X-axis.

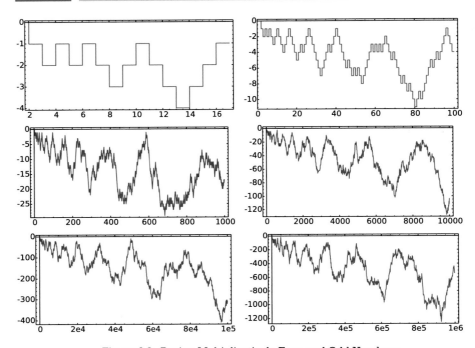

Figure 6.3. Racing Multiplicatively Even and Odd Numbers.

A *negative* response to this question – i.e., a proof that any plot as drawn in Figure 6.3 never crosses the X-axis – would imply the Riemann Hypothesis! In contrast to the list of previous questions, the answer to this question is known[2]: alas, there is such an X. In 1960, Lehman showed that for $X = 906,400,000$ there are 708 more multiplicatively even numbers up to X than multiplicatively odd numbers (Tanaka found in 1980 that the smallest X such that there are more multiplicative even than odd numbers is $X = 906,150,257$).

These are questions that have been asked about primes (and we could give bushels more[3]), questions expressible in simple vocabulary, that we can't answer today. We have been studying numbers for over two millennia and yet we are indeed in the infancy of our understanding.

We'll continue our discussion by returning to the simplest counting question about prime numbers.

---

[2] For more details, see P. Borwein, "Sign changes in sums of the Liouville Function" and the nice short paper of Norbert Wiener "Notes on Polya's and Turan's hypothesis concerning Liouville's factor" (page 765 of volume II of Wiener's Collected Works); see also: G. Pólya "Verschiedene Bemerkungen zur Zahlentheorie," *Jahresbericht der Deutschen Mathematiker-Vereinigung,* **28** (1919) 31–40.

[3] See, e.g., Richard Guy's book *Unsolved Problems in Number Theory* (2004).

7 How Many Primes are There?

	2	3		5		7				11		13	
		17		19				23					
29		31						37				41	
43				47						53			
		59		61						67			
71		73						79				83	
				89								97	
		101		103				107		109			
113													
127				131						137		139	
								149		151			
		157						163				167	
169				173						179		181	
								191		193			

Figure 7.1. Sieving primes up to 200

Slow as we are to understand primes, at the very least we can try to count them. You can see that there are 10 primes less than 30, so you might encapsulate this by saying that the chances that a number less than 30 is prime is 1 in 3. This frequency does not persist, though; here is some more data: There are 25 primes less than 100 (so 1 in 4 numbers up to 100 are prime), there are 168 primes less than a thousand (so we might say that among the numbers less than a thousand the chances that one of them is prime is roughly 1 in 6).

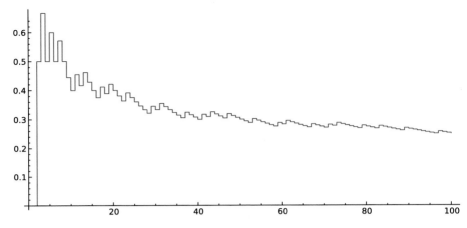

Figure 7.2. Graph of the proportion of primes up to X for each integer $X \leq 100$

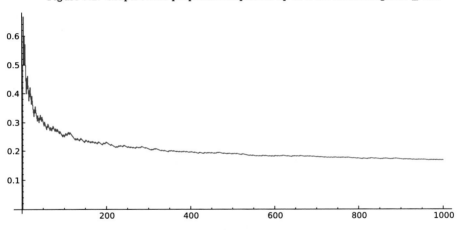

Figure 7.3. Proportion of primes for X up to 1,000

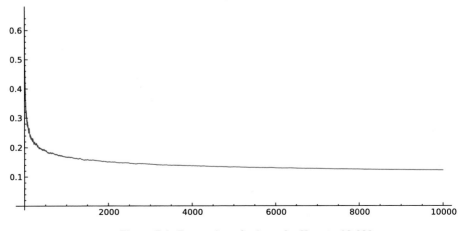

Figure 7.4. Proportion of primes for X up to 10,000

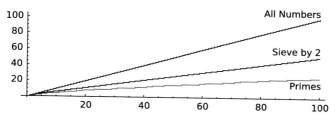

Figure 7.5. Sieving by removing multiples of 2 up to 100

There are 78,498 primes less than a million (so we might say that the chances that a random choice among the first million numbers is prime have dropped to roughly 1 in 13).

There are 455,052,512 primes less than ten billion; i.e., 10,000,000,000 (so we might say that the chances are down to roughly 1 in 22).

Primes, then, seem to be thinning out. We return to the sifting process we carried out earlier, and take a look at a few graphs, to get a sense of why that might be so. There are a 100 numbers less than or equal to 100, a thousand numbers less than or equal to 1000, etc.: the graph in Figure 7.5 that looks like a regular staircase, each step the same length as each riser, climbing up at, so to speak, a 45 degree angle, counts all numbers up to and including X.

Following Eratosthenes, we have sifted those numbers, to pan for primes. Our first move was to throw out roughly half the numbers (the even ones!) after the number 2. The cross-hatched bar graph in this figure that is, with one hiccup, a regular staircase climbing at a smaller angle, each step twice the length of each riser, illustrates the numbers that are left after one pass through Eratosthenes' sieve, which includes, of course, all the primes. So, the chances that a number bigger than 2 is prime is *at most* 1 in 2. Our second move was to throw out a good bunch of numbers bigger than 3. So, the chances that a number bigger than 3 is prime is going to be even less. And so it goes: with each move in our sieving process, we are winnowing the field more extensively, reducing the chances that the later numbers are prime.

The red curve in these figures actually counts the primes: it is the beguilingly irregular *staircase of primes*. Its height above any number X on the horizontal line records the number of primes less than or equal to X, the accumulation of primes up to X. Refer to this number as $\pi(X)$. So $\pi(2) = 1$, $\pi(3) = 2$,

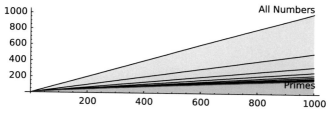

Figure 7.6. Sieving for primes up to 1000

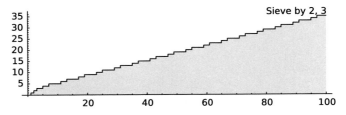

Figure 7.7. Sieving out multiples of 2 and 3

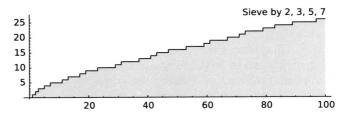

Figure 7.8. Sieving out multiples of 2, 3, 5, and 7

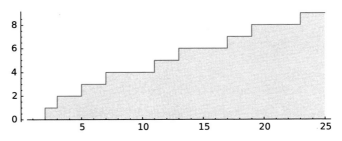

Figure 7.9. Staircase of primes up to 25

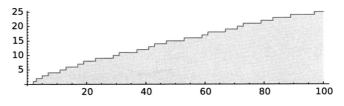

Figure 7.10. Staircase of primes up to 100

$\pi(30) = 10$; of course, we could plot a few more values of $\pi(X)$, like $\pi(\text{ten billion}) = 455,052,512$.

Let us accompany Eratosthenes for a few further steps in his sieving process. Figure 7.7 contains a graph of all whole numbers up to 100 after we have removed the even numbers greater than 2, and the multiples of 3 greater than 3 itself.

From this graph you can see that if you go "out a way" the likelihood that a number is a prime is less than 1 in 3. Figure 7.8 contains a graph of what Eratosthenes sieve looks like up to 100 after sifting 2, 3, 5, and 7.

This data may begin to suggest to you that as you go further and further out on the number line the percentage of prime numbers among all whole numbers tends towards 0% (it does).

To get a sense of how the primes accumulate, we will take a look at the staircase of primes for $X = 25$ and $X = 100$ in Figures 7.9 and 7.10.

8 Prime Numbers Viewed from a Distance

The striking thing about these figures is that as the numbers get large enough, the jagged accumulation of primes, those quintessentially discrete entities, becomes smoother and smoother to the eye. How strange and wonderful to watch, as our viewpoint zooms out to larger ranges of numbers, the accumulation of primes taking on such a smooth and elegant shape.

But don't be fooled by the seemingly smooth shape of the curve in the last figure above: it is just as faithful a reproduction of the staircase of primes as the typographer's art can render, for there are thousands of tiny steps and risers in this curve, all hidden by the thickness of the print of the drawn curve in the figure. It is already something of a miracle that we can approximately describe the build-up of primes, somehow, using a *smooth curve*. But *what* smooth curve?

That last question is *not* rhetorical. If you draw a curve with chalk on the blackboard, this can signify a myriad of smooth (mathematical) curves

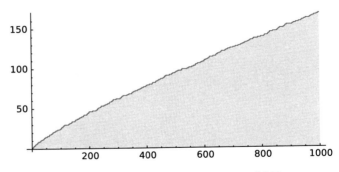

Figure 8.1. Staircases of primes up to 1,000

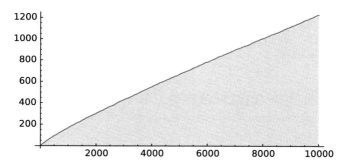

Figure 8.2. Staircases of primes up to 10,000

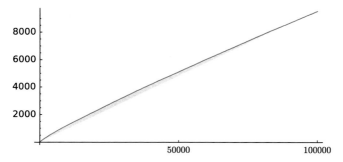

Figure 8.3. Primes up to 100,000

all encompassed within the thickness of the chalk-line, all – if you wish – reasonable approximations of one another. So, there are many smooth curves that fit the chalk-curve. With this warning, but very much fortified by the data of Figure 8.3, let us ask: *what is a smooth curve that is a reasonable approximation to the staircase of primes?*

9 Pure and Applied Mathematics

Mathematicians seems to agree that, loosely speaking, there are two types of mathematics: *pure* and *applied*. Usually – when we judge whether a piece of mathematics is pure or applied – this distinction turns on whether or not the math has application to the "outside world," i.e., that *world* where bridges are built, where economic models are fashioned, where computers churn away on the Internet (for only then do we unabashedly call it *applied math*), or whether the piece of mathematics will find an important place within the context of mathematical theory (and then we label it *pure*). Of course, there is a great overlap (as we will see later, Fourier analysis plays a major role both in data compression and in pure mathematics).

Moreover, many questions in mathematics are "hustlers" in the sense that, at first view, what is being requested is that some simple task be done (e.g., the question raised in this book, *to find a smooth curve that is a reasonable approximation to the staircase of primes*). And only as things develop is it discovered that there are payoffs in many unexpected directions, some of these payoffs being genuinely applied (i.e., to the practical world), some of these payoffs being pure (allowing us to strike behind the mask of the mere appearance of the mathematical situation, and get at the hidden fundamentals that actually govern the phenomena), and some of these payoffs defying such simple classification, insofar as they provide powerful techniques in other branches of mathematics. The Riemann Hypothesis – even in its current unsolved state – has already shown itself to have all three types of payoff.

The particular issue before us is, in our opinion, twofold, both applied, and pure: can we curve-fit the "staircase of primes" by a well approximating smooth curve given by a simple analytic formula? The story behind this alone is marvelous, has a cornucopia of applications, and we will be telling it below.

But our curiosity here is driven by a question that is pure, and less amenable to precise formulation: are there mathematical concepts at the root of, and more basic than (and "prior to," to borrow Aristotle's use of the phrase) *prime numbers* – concepts that account for the apparent complexity of the nature of primes?

10 | A Probabilistic First Guess

Figure 10.1. Carl Friedrich Gauss (1824–1908). Courtesy of the Smithsonian Libraries, Washington, D.C

The search for such approximating curves began, in fact, two centuries ago when Carl Friedrich Gauss defined a certain beautiful curve that, experimentally, seemed to be an exceptionally good fit for the staircase of primes.

Let us denote Gauss's curve $G(X)$; it has an elegant simple formula comprehensible to anyone who has had a tiny bit of calculus. If you make believe that the chances that a number X is a prime is inversely proportional to the number of digits of X you might well hit upon Gauss's curve. That is,

$G(X)$ is roughly proportional to $\dfrac{X}{\text{the number of digits of } X}.$

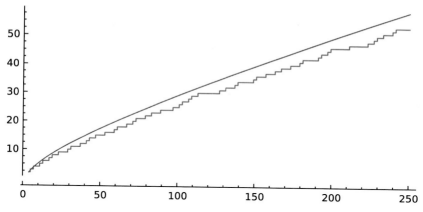

Figure 10.2. Plot of $\pi(X)$ and Gauss's smooth curve $G(X)$

But to describe Gauss's guess precisely we need to discuss the *natural loga-rithm*[1] "log(X)" which is an elegant smooth function of a real number X that is roughly proportional to the number of digits of the whole number part of X.

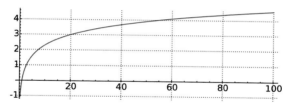

Figure 10.3. Plot of the natural logarithm $\log(X)$

Euler's famous constant $e = 2.71828182\ldots$, which is the limit of the sequence

$$\left(1 + \frac{1}{2}\right)^2, \left(1 + \frac{1}{3}\right)^3, \left(1 + \frac{1}{4}\right)^4, \ldots,$$

is used in the definition of log:

$$A = \log(X) \text{ is the number } A \text{ for which } e^A = X.$$

Before electronic calculators, logarithms were frequently used to speed up calculations, since logarithms translate difficult multiplication problems into easier addition problems which can be done mechanically. Such calculations use that the logarithm of a product is the sum of the logarithms of the factors; that is,

$$\log(XY) = \log(X) + \log(Y).$$

---

[1] In advanced mathematics, "common" logarithms are sufficiently uncommon that "log" almost always denotes natural log and the notation $\ln(X)$ is not used.

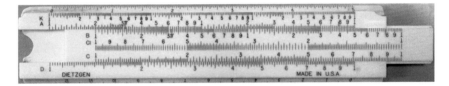

Figure 10.4. A slide rule computes $2X$ by using that $\log(2X) = \log(2) + \log(X)$. Photo by William Stein

In Figure 10.4 the numbers printed (on each of the slidable pieces of the rule) are spaced according to their logarithms, so that when one slides the rule arranging it so that the printed number X on one piece lines up with the printed number 1 on the other, we get that for every number Y printed on the first piece, the printed number on the other piece that is aligned with it is the product XY; in effect the "slide" adds $\log(X)$ to $\log(Y)$ giving $\log(XY)$.

Unter	gibtes Primzahlen	Integral $\int \frac{dn}{\log n}$	Differ	Ihre Formel	Abweich.
500 000	41 556	41606,4	+50,4	41596,9	+40,9
1000 000	78 501	78627,5	+126,5	78672,7	+171,7
1500 000	114 112	114263,1	+151,1	114374,0	+264,0
2000 000	148883	149054,8	+171,8	149233,0	+350,0
2500 000	183016	183245,0	+229,0	183495,1	+479,1
3000 000	216745	216970,6	+225,6	217308,5	+563,6

Dass Legendre sich auch mit diesem Gegenstande beschäftigt hat, war mir nicht bekannt; auf Veranlassung Ihres Briefes habe ich in seine Theorie des Nombres nachgesehen, und in der zweiten Ausgabe einige darauf bezügliche Seiten gefunden, die ich früher übersehen (oder seitdem verges-
sen) haben muß. Legendre gebrauchs die Formel

$$\frac{n}{\log n - A}$$

Figure 10.5. A Letter of Gauss. Courtesy of SUB Göttingen

In 1791, when Gauss was 14 years old, he received a book that contained logarithms of numbers up to 7 digits and a table of primes up to 10,009. Years later, in a letter written in 1849 (see Figure 10.5), Gauss claimed that as early as 1792 or 1793 he had already observed that the density of prime numbers over intervals of numbers of a given rough magnitude X seemed to average $1/\log(X)$.

Very *very* roughly speaking, this means that *the number of primes up to X is approximately X divided by twice the number of digits of X.* For example, the number of primes less than 99 should be roughly

$$\frac{99}{2 \times 2} = 24.75 \approx 25,$$

which is pretty amazing, since the correct number of primes up to 99 is 25. The number of primes up to 999 should be roughly

$$\frac{999}{2 \times 3} = 166.5 \approx 167,$$

which is again close, since there are 168 primes up to 1000. The number of primes up to 999,999 should be roughly

$$\frac{999999}{2 \times 6} = 83333.25 \approx 83,333,$$

which is close to the correct count of 78,498.

Gauss guessed that the expected number of primes up to X is approximated by the area under the graph of $1/\log(X)$ from 2 to X (see Figure 10.6). The area under $1/\log(X)$ up to $X = 999,999$ is $78,626.43\ldots$, which is remarkably close to the correct count 78,498 of the primes up to 999,999.

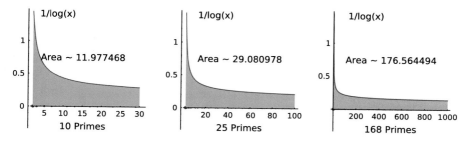

Figure 10.6. The expected tally of the number of primes $\leq X$ is approximated by the area underneath the graph of $1/\log(X)$ from 1 to X.

Gauss was an inveterate computer: he wrote in his 1849 letter that there are 216,745 prime numbers less than three million. This is wrong: the actual number of these primes is 216,816. Gauss's curve $G(X)$ predicted that there would be 216,970 primes – a miss, Gauss thought, by

$$225 = 216970 - 216745.$$

But actually he was closer than he thought: the prediction of the curve $G(X)$ missed by a mere $154 = 216970 - 216816$. Gauss's computation brings up two queries: will this spectacular "good fit" continue for arbitrarily large numbers? and, the (evidently prior) question: what counts as a good fit?

11 What is a "Good Approximation"?

If you are trying to estimate a number, say, around ten thousand, and you get it right to within a hundred, let us celebrate this kind of accuracy by saying that you have made an approximation with *square-root error* ($\sqrt{10,000} = 100$). Of course, we should really use the more clumsy phrase "an approximation with at worst *square-root error*." Sometimes we'll simply refer to such approximations as *good approximations*. If you are trying to estimate a number in the millions, and you get it right to within a thousand, let's agree that – again – you have made an approximation with *square-root error* ($\sqrt{1,000,000} = 1,000$). Again, for short, call this a *good approximation*. So, when Gauss thought his curve missed by 226 in estimating the number of primes less than three million, it was well within the margin we have given for a "good approximation."

More generally, if you are trying to estimate a number that has D digits and you get it almost right, but with an error that has no more than, roughly, half that many digits, let us say, again, that you have made an approximation with *square-root error* or synonymously, a *good approximation*.

This rough account almost suffices for what we will be discussing below, but to be more precise, the specific *gauge of accuracy* that will be important to us is not for a mere *single* estimate of a *single* error term,

$$\text{Error term} = \text{Exact value} - \text{Our "good approximation"}$$

but rather for *infinite sequences* of estimates of error terms. Generally, if you are interested in a numerical quantity $q(X)$ that depends on the real number parameter X (e.g., $q(X)$ could be $\pi(X)$, "the number of primes $\leq X$") and if you have an explicit candidate "approximation," $q_{\text{approx}}(X)$, to this quantity, let us say that $q_{\text{approx}}(X)$ is **essentially a square-root accurate approximation to** $q(X)$ if for *any* given exponent greater than 0.5 (you choose it: 0.501, 0.5001,

0.50001, ... for example) and for large enough X – where the phrase "large enough" depends on your choice of exponent – the **error term** – i.e., the difference between $q_{approx}(X)$ and the *true* quantity, $q(X)$, is, in absolute value, less than X raised to that exponent (e.g. $< X^{0.501}$, $< X^{0.5001}$, etc.). Readers who know calculus and wish to have a technical formulation of this definition of *good approximation* might turn to the endnote [7] for a precise statement.

If you found the above confusing, don't worry: again, a square-root accurate approximation is one in which at least roughly half the digits are correct.

Remark 11.1

To get a feel for how basic the notion of *approximation to data being square root close to the true values of the data* is – and how it represents the "gold standard" of accuracy for approximations, consider this fable.

Imagine that the devil had the idea of saddling a large committee of people with the task of finding values of $\pi(X)$ for various large numbers X. This he did in the following manner, having already worked out which numbers are prime himself. Since the devil is, as everyone knows, *in the details,* he has made no mistakes: his work is entirely correct. He gives each committee member a copy of the list of all *prime numbers* between 1 and one of the large numbers X in which he was interested. Now each committee member would count the number of primes by doing nothing more than considering each number, in turn, on their list and tallying them up, much like a canvasser counting votes; the committee members needn't even know that these numbers are prime, they just think of these numbers as items on their list. But since they are human, they will indeed be making mistakes, say 1% of the time. Assume further that it is just as likely for them to make the mistake of undercounting or overcounting. If many people are engaged in such a pursuit, some of them might overcount $\pi(X)$; some of them might undercount it. The average error (overcounted or undercounted) would be proportional to $\sqrt{X}$.

In the next chapter we'll view these undercounts and overcounts as analogous to a random walk.

12 Square Root Error and Random Walks

To take a random walk along a (straight) east–west path you would start at your home base, but every minute, say, take a step along the path, each step being of the same length, but randomly either east or west. After X minutes, how far are you from your home base?

The answer to this *cannot* be a specific number, precisely because you're making a random decision that affects that number for each of the X minutes of your journey. It is more reasonable to ask a statistical version of that question. Namely, if you took *many* random walks X minutes long, then – on the average – how far would you be from your home base? The answer, as is illustrated by the figures below, is that the average distance you will find yourself from home base after (sufficiently many of) these excursions is proportional to $\sqrt{X}$. (In fact, the average is equal to $\sqrt{\frac{2}{\pi}} \cdot \sqrt{X}$.)

To connect this with the committee members' histories of errors, described in the fable in Chapter 11, imagine every error (undercount or overcount by 1) the committee member makes, as a "step" East for undercount and West for overcount. Then if such errors were made, at a constant frequency over the duration of time spent counting, and if the over and undercounts were equally likely and random, then one can model the committee members' computational accuracy by a random walk. It would be – in the terminology we have already discussed – no better than *square-root accurate*; it would be subject to *square-root error*.

To get a real sense of how constrained random walks are by this "square-root law," here are a few numerical experiments of random walks. The left-hand squiggly (blue) graphs in Figures 12.1–12.4 below are computer-obtained random walk trials (three, ten, a hundred, and a thousand random walks). The blue curve in the right-hand graphs of those four figures is the average distance

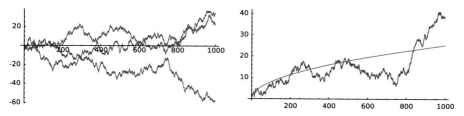

Figure 12.1. Three Random Walks

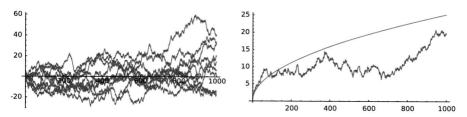

Figure 12.2. Ten Random Walks

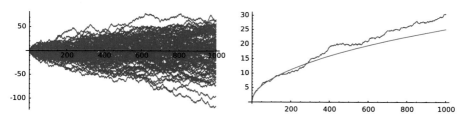

Figure 12.3. One Hundred Random Walks

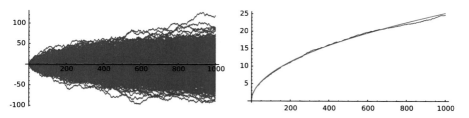

Figure 12.4. One Thousand Random Walks

from home-base of the corresponding (three, ten, a hundred, and a thousand) random walks. The red curve in each figure below is the graph of the quantity $\sqrt{\frac{2}{\pi}} \cdot \sqrt{X}$ over the X-axis. As the number of random walks increases, the red curve better and better approximates the average distance.

13 What is Riemann's Hypothesis? (First Formulation)

Recall from Chapter 10 that a rough guess for an approximation to $\pi(X)$, the number of primes $\leq X$, is given by the function $X/\log(X)$. Recall, as well, that a refinement of that guess, offered by Gauss, stems from this curious thought: the "probability" that a number N is a prime is proportional to the reciprocal of its number of digits; more precisely, the probability is $1/\log(N)$. This would lead us to guess that the approximate value of $\pi(X)$ would be the area of the region from 2 to X under the graph of $1/\log(X)$, a quantity sometimes referred to as $\mathrm{Li}(X)$. "Li" (pronounced $\overline{\mathrm{Li}}$, so the same as "lie" in "lie down") is short for *logarithmic integral*, because the area of the region from 2 to X under $1/\log(X)$ is (by definition) the *integral* $\int_2^X 1/\log(t)\,dt$.

Figure 13.1 contains a graph of the three functions $\mathrm{Li}(X)$, $\pi(X)$, and $X/\log X$ for $X \leq 200$. But data, no matter how impressive, may be deceiving (as we learned in Chapter 6). If you think that the three graphs never cross for all large

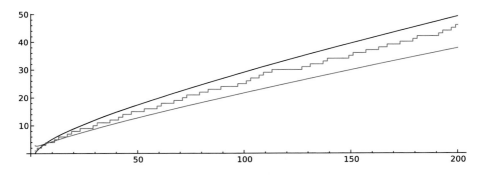

Figure 13.1. Plots of $\mathrm{Li}(X)$ (top), $\pi(X)$ (in the middle), and $X/\log(X)$ (bottom)

values of X, and that we have the simple relationship $X/\log(X) < \pi(X) < \mathrm{Li}(X)$ for large X, read `http://en.wikipedia.org/wiki/Skewes'_number`.

It is a *major challenge* to evaluate $\pi(X)$ for large values of X. For example, let $X = 10^{24}$. Then (see [8]) we have:

$$\pi(X) = 18{,}435{,}599{,}767{,}349{,}200{,}867{,}866$$

$$\mathrm{Li}(X) = 18{,}435{,}599{,}767{,}366{,}347{,}775{,}143.10580\ldots$$

$$X/(\log(X) - 1) = 18{,}429{,}088{,}896{,}563{,}917{,}716{,}962.93869\ldots$$

$$\mathrm{Li}(X) - \pi(X) = \qquad\qquad 17{,}146{,}907{,}277.105803\ldots$$

$$\sqrt{X} \cdot \log(X) = \qquad\qquad 55{,}262{,}042{,}231{,}857.096416\ldots$$

Note that several of the left-most digits of $\pi(X)$ and $\mathrm{Li}(X)$ are the same (as indicated in red), a point we will return to on page 48.

More fancifully, we can think of the error in this approximation to $\pi(X)$, i.e., $|\mathrm{Li}(X) - \pi(X)|$, (the absolute value) of the difference between $\mathrm{Li}(X)$ and $\pi(X)$, as (approximately) the result of a walk having roughly X steps where you move by the following rule: go east by a distance of $1/\log N$ feet if N is not a prime and west by a distance of $1 - \frac{1}{\log N}$ feet if N is a prime. Your distance, then, from home base after X steps is approximately $|\mathrm{Li}(X) - \pi(X)|$ feet.

We have no idea if this picture of things resembles a truly *random walk* but at least it makes it reasonable to ask the question: is $\mathrm{Li}(X)$ *essentially a square root approximation* to $\pi(X)$? Our first formulation of Riemann's Hypothesis says yes:

The Riemann Hypothesis (first formulation)

For any real number X the number of prime numbers less than X is approximately $\mathrm{Li}(X)$ and this approximation is essentially square root accurate (see [9]).

14 The Mystery Moves to the Error Term

Let's think of what happens when you have a *mysterious quantity* (say, a function of a real number X) you wish to understand. Suppose you manage to approximate that quantity by an easy to understand expression – which we'll call the "dominant term" – that is also simple to compute, but only approximates your mysterious quantity. The approximation is not exact; it has a possible error, which happily is significantly smaller than the size of the dominant term. "Dominant" here just means exactly that: it is of size significantly larger than the error of approximation.

Mysterious quantity$(X) =$ *Simple, but dominant quantity*$(X) +$ *Error*(X).

If all you are after is a general estimate of *size* your job is done. You might declare victory and ignore *Error*(X) as being – in size – negligible. But if you are interested in the deep structure of your *mysterious quantity* perhaps all you have done is to manage to transport most questions about it to *Error*(X). In conclusion, *Error*(X) is now your new mysterious quantity.

Returning to the issue of $\pi(X)$ (our *mysterious quantity*) and Li(X) (our *dominant term*) the first formulation of the Riemann Hypothesis (as in Chapter 13 above) puts the spotlight on the *Error term* $|$Li$(X) - \pi(X)|$, which therefore deserves our scrutiny, since – after all – we're not interested in merely counting the primes: we want to understand as much as we can of their structure.

To get a feel for this error term, we shall smooth it out a bit, and look at a few of its graphs.

15 Cesàro Smoothing

Often when you hop in a car and reset the trip counter, the car will display information about your trip. For instance, it might show you the average speed up until now, a number that is "sticky", changing much less erratically than your actual speed, and you might use it to make a rough estimate of how long until you will reach your destination. Your car is computing the *Cesàro smoothing* of your speed. We can use this same idea to better understand the behavior of other things, such as the sums appearing in the previous chapter.

Figure 15.1. The "Average Vehicle Speed," as displayed on the dashboard of the 2013 Camaro SS car that one of us drove during the writing of this. Photo by William Stein

Suppose you are trying to say something intelligent about the behavior of a certain quantity that varies with time in what seems to be a somewhat erratic, volatile pattern. Call the quantity $f(t)$ and think of it as a function defined for positive real values of "time" t. The natural impulse might be to take some sort

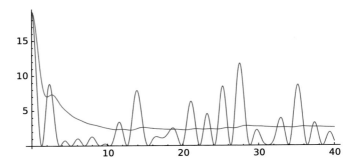

Figure 15.2. The red plot is the Cesàro smoothing of the blue plot

of "average value"[1] of $f(t)$ over a time interval, say from 0 to T. This would indeed be an intelligent thing to do if this average stabilized and depended relatively little on the interval of time over which one is computing this average, e.g., if that interval were large enough. Sometimes, of course, these averages themselves are relatively sensitive to the times T that are chosen, and don't stabilize. In such a case, it usually pays to consider *all* these averages as your chosen T varies, as a function (of T) in its own right[2]. This new function $F(T)$, called the **Cesàro Smoothing** of the function $f(t)$, is a good indicator of certain eventual trends of the original function $f(t)$ that is less visible directly from the function $f(t)$. The effect of passing from $f(t)$ to $F(T)$ is to "smooth out" some of the volatile local behavior of the original function, as can be seen in the Figure 15.2.

---

[1] For readers who know calculus: that average would be $\frac{1}{T} \int_0^T f(t)dt$.
[2] For readers who know calculus: this is

$$F(T) := \frac{1}{T} \int_0^T f(t)dt.$$

16 A View of $|\text{Li}(X) - \pi(X)|$

Returning to our mysterious error term, consider Figure 16.1 where the volatile blue curve in the middle is $\text{Li}(X) - \pi(X)$, its Cesàro smoothing is the red relatively smooth curve on the bottom, and the curve on the top is the graph of $\sqrt{\frac{2}{\pi}} \cdot \sqrt{X/\log(X)}$ over the range of $X \leq 250,000$.

Data such as this graph can be revealing and misleading at the same time. For example, the volatile blue graph (of $\text{Li}(X) - \pi(X)$) *seems* to be roughly sandwiched between two rising functions, but this will not continue for all values of X. The Cambridge University mathematician, John Edensor Littlewood (see Figure 16.2), proved in 1914 that there *exists* a real number X for which the quantity $\text{Li}(X) - \pi(X)$ vanishes, and then for slightly larger X crosses over into negative values. This theorem attracted lots of attention at the time (and continues to do so) because of the inaccessibility of achieving good estimates (upper or lower bounds) for the first such number. That "first" X for which

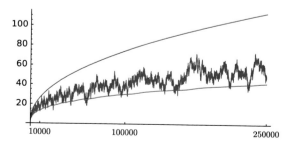

Figure 16.1. $\text{Li}(X) - \pi(X)$ (blue middle), its Cesàro smoothing (red bottom), and $\sqrt{\frac{2}{\pi}} \cdot \sqrt{X/\log(X)}$ (top), all for $X \leq 250,000$

Figure 16.2. John Edensor Littlewood (1885–1977). Courtesy of Trinity College Library

$\text{Li}(X) = \pi(X)$ is called **Skewes Number** in honor of the South African mathematician (a student of Littlewood) Stanley Skewes, who (in 1933) gave the first – fearfully large – upper bound for that number (conditional on RH!). Despite a steady stream of subsequent improvements, we currently can only locate Skewes Number as being in the range:

$$10^{14} \leq \text{Skewes Number} < 10^{317},$$

and it is proven that at some values of X fairly close to the upper bound 10^{317} $\pi(X)$ is greater than $\text{Li}(X)$. So the trend suggested in Figure 16.1 will not continue indefinitely.

17 The Prime Number Theorem

Take a look at Figure 13.1 again. All three functions, $\text{Li}(X)$, $\pi(X)$ and $X/\log(X)$ are "going to infinity with X" (this means that for any real number R, for all sufficiently large X, the values of these functions at X exceed R).

Are these functions "going to infinity" at *the same rate*?

To answer such a question, we have to know what we mean by *going to infinity at the same rate*. So, here's a definition. Two functions, $A(X)$ and $B(X)$, that each go to infinity will be said to **go to infinity at the same rate** if their *ratio*

$$A(X)/B(X)$$

tends to 1 as X goes to infinity.

If for example two functions, $A(X)$ and $B(X)$ that take positive whole number values, have the same number of digits for large X and if, for any number you give us, say: a million (or a billion, or a trillion) and if X is large enough, then the "leftmost" million (or billion, or trillion) digits of $A(X)$ and $B(X)$ are the same, then $A(X)$ and $B(X)$ *go to infinity at the same rate*. For example,

$$\frac{A(X)}{B(X)} = \frac{28106759718374352510561175 5423}{28106759716136151152776629 4585} = 1.00000000007963213762060\ldots$$

While we're defining things, let us say that two functions, $A(X)$ and $B(X)$, that each go to infinity **go to infinity at similar rates** if there are two positive constants c and C such that for X sufficiently large the *ratio*

$$A(X)/B(X)$$

is greater than c and smaller than C.

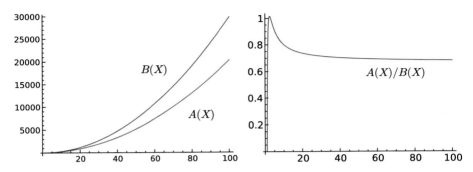

Figure 17.1. The polynomials $A(X) = 2X^2 + 3X - 5$ (bottom) and $B(X) = 3X^2 - 2X + 1$ (top) go to infinity at similar rates

For example, two polynomials in X with positive leading coefficients *go to infinity at the same rate* if and only if they have the same degrees and the same leading coefficient; they *go to infinity at similar rates* if they have the same degree. See Figures 17.1 and 17.2.

Now a theorem from elementary calculus tells us that the ratio of $\mathrm{Li}(X)$ to $X/\log(X)$ tends to 1 as X gets larger and larger. That is – using the definition we've just introduced – $\mathrm{Li}(X)$ and $X/\log(X)$ go to infinity at the same rate (see [10]).

Recall (on page 41 above in Chapter 13) that if $X = 10^{24}$, the left-most twelve digits of $\pi(X)$ and $\mathrm{Li}(X)$ are the same: both numbers start $18,435,599,767,3\ldots$. Well, that's a good start. Can we guarantee that for X large enough, the "left-most" million (or billion, or trillion) digits of $\pi(X)$ and $\mathrm{Li}(X)$ are the same, i.e., that these two functions go to infinity at the same rate?

The Riemann Hypothesis, as we have just formulated it, would tell us that the *difference* between $\mathrm{Li}(X)$ and $\pi(X)$ is pretty small in comparison with the size of X. This information would imply (but would be *much* more precise information than) the statement that the *ratio* $\mathrm{Li}(X)/\pi(X)$ tends to 1, i.e., that $\mathrm{Li}(X)$ and $\pi(X)$ go to infinity at the same rate.

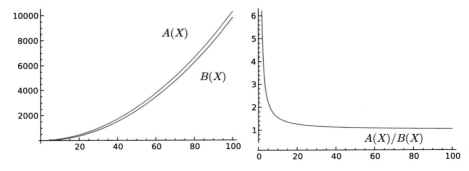

Figure 17.2. The polynomials $A(X) = X^2 + 3X - 5$ (top) and $B(X) = X^2 - 2X + 1$ (bottom) go to infinity at the same rate

Bei dieser Untersuchung diente mir als Ausgangspunkt die von Euler gemachte Bemerkung, dass das Product

$$\prod \frac{1}{1-\frac{1}{p^s}} = \sum \frac{1}{n^s},$$

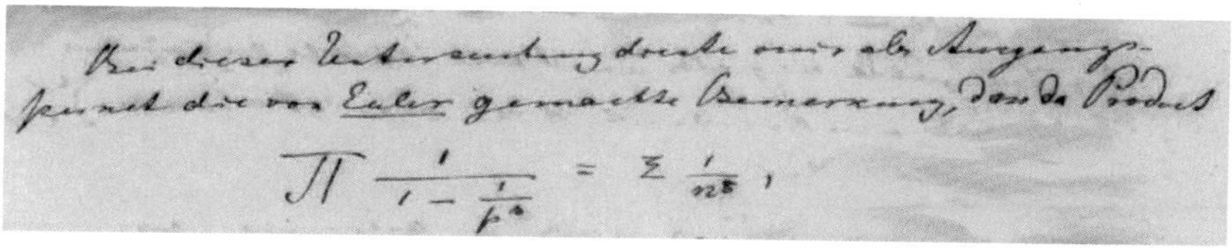

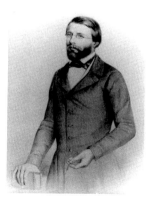

Figure 17.3. From a manuscript of Riemann's 1859 paper written by a contemporary of Riemann (possibly the mathematician Alfred Clebsch). Courtesy of SUB Göttingen

This last statement gives, of course, a far less precise relationship between $\mathrm{Li}(X)$ and $\pi(X)$ than the Riemann Hypothesis (once it is proved!) would give us. The advantage, though, of the less precise statement is that it is currently known to be true, and – in fact – has been known for over a century. It goes under the name of

The Prime Number Theorem: $\mathrm{Li}(X)$ and $\pi(X)$ go to infinity at the same rate.

Since $\mathrm{Li}(X)$ and $X/\log(X)$ go to infinity at the same rate, we could equally well have expressed the "same" theorem by saying:

The Prime Number Theorem: $X/\log(X)$ and $\pi(X)$ go to infinity at the same rate.

This fact is a very hard-won piece of mathematics! It was proved in 1896 independently by Jacques Hadamard and Charles de la Vallée Poussin.

A milestone in the history leading up to the proof of the Prime Number Theorem is the earlier work of Pafnuty Lvovich Chebyshev (see `http://en.wikipedia.org/wiki/Chebyshev_function`) showing that (to use the terminology we introduced) $X/\log(X)$ and $\pi(X)$ go to infinity at similar rates.

The elusive Riemann Hypothesis, however, is much deeper than the Prime Number Theorem, and takes its origin from some awe-inspiring, difficult to interpret, lines in Bernhard Riemann's magnificent 8-page paper, "On the number of primes less than a given magnitude," published in 1859 (see [11]).

Figure 17.4. Bernhard Riemann (1826–1866) copyright © Archives of the Mathematisches Forschungsinstitut Oberwolfach

Riemann's hypothesis, as it is currently interpreted, turns up as relevant, as a key, again and again in different parts of the subject: if you accept it as *hypothesis* you have an immensely powerful tool at your disposal: a mathematical magnifying glass that sharpens our focus on number theory. But it also has a wonderful protean quality – there are many ways of formulating it, any of these formulations being provably equivalent to any of the others.

The Riemann Hypothesis remains unproved to this day, and therefore is "only a hypothesis," as Osiander said of Copernicus's theory, but one for which we have overwhelming theoretical and numerical evidence in its support. It is the kind of conjecture that contemporary Dutch mathematician Frans Oort might label a *suffusing conjecture* in that it has unusually broad implications: many, many results are now known to follow, if the conjecture, familiarly known as RH, is true. A proof of RH would, therefore, fall into the *applied* category, given our discussion above in Chapter 9. But however you classify RH, it is a central concern in mathematics to find its proof (or, a counter-example!). RH is one of the weightiest statements in all of mathematics.

18 The Information Contained in the Staircase of Primes

We have borrowed the phrase "staircase of primes" from the popular book *The Music of the Primes* by Marcus du Sautoy, for we feel that it captures the sense that there is a deeply hidden architecture to the graphs that compile the number of primes (up to N) and also because – in a bit – we will be tinkering with this carpentry. Before we do so, though, let us review in Figure 18.1 what this staircase looks like for different ranges.

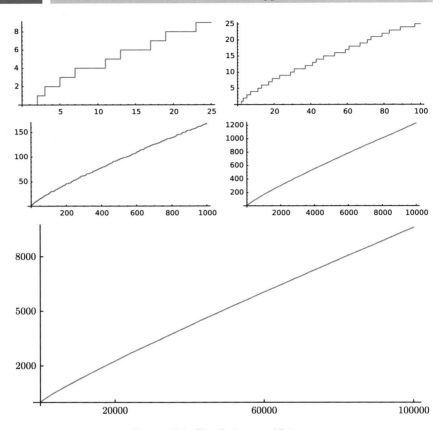

Figure 18.1. The Staircase of Primes

The mystery of this staircase is that the *information* contained within it is – in effect – the full story of where the primes are placed. This story seems to elude any simple description. Can we "tinker with" this staircase without destroying this valuable information?

19 | Tinkering with the Carpentry of the Staircase of Primes

For starters, notice that all the (vertical) *risers* of this staircase (Figure 18.1 above) have unit height. That is, they contain no numerical information except for their placement on the x-axis. So, we could distort our staircase by changing (in any way we please) the height of each riser; and as long as we haven't brought new risers into – or old risers out of – existence, and have not modified their position over the x-axis, we have retained all the information of our original staircase.

A more drastic-sounding thing we could do is to judiciously add new steps to our staircase. At present, we have a step at each prime number p, and no step anywhere else. Suppose we built a staircase with a new step not only at $x = p$ for p each prime number but also at $x = 1$ and $x = p^n$ where p^n runs through all powers of prime numbers as well. Such a staircase would have, indeed, many more steps than our original staircase had, but, nevertheless, would retain much of the quality of the old staircase: namely it contains within it the full story of the placement of primes *and their powers*.

A final thing we can do is to perform a distortion of the x-axis (elongating or shortening it, as we wish) in any specific way, as long as we can perform the inverse process, and "undistort" it if we wish. Clearly such an operation may have mangled the staircase, but hasn't destroyed information irretrievably.

We shall perform all three of these kinds of operations eventually, and will see some great surprises as a result. But for now, we will perform distortions only of the first two types. We are about to build a new staircase that retains the precious information we need, but is constructed according to the following architectural plan.

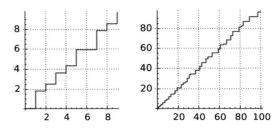

Figure 19.1. The newly constructed staircase that counts prime powers

- We first build a staircase that has a new step precisely at $x = 1$, and $x = p^n$ for every *prime power* p^n with $n \geq 1$. That is, there will be a new step at $x = 1, 2, 3, 4, 5, 7, 8, 9, 11, \ldots$
- Our staircase starts on the ground at $x = 0$ and the height of the riser of the step at $x = 1$ will be $\log(2\pi)$. The height of the riser of the step at $x = p^n$ will not be 1 (as was the height of all risers in the old staircase of primes) but rather: the step at $x = p^n$ will have the height of its riser equal to $\log p$. So for the first few steps listed in the previous item, the risers will be of height $\log(2\pi), \log 2, \log 3, \log 2, \log 5, \log 7, \log 2, \log 3, \log 11, \ldots$ Since $\log(p) > 1$, these vertical dimensions lead to a steeper ascent but no great loss of *information*.

  Although we are not quite done with our architectural work, Figure 19.1 shows what our new staircase looks like, so far.

Notice that this new staircase looks, from afar, as if it were nicely approximated by the 45 degree straight line, i.e., by the simple function X. In fact, we have – by this new architecture – a second *equivalent* way of formulating Riemann's hypothesis. For this, let $\psi(X)$ denote the function of X whose graph is depicted in Figure 19.1 (see [12]).

The Riemann Hypothesis (second formulation)

This new staircase is essentially square root close to the 45 degree straight line; i.e., the function $\psi(X)$ is essentially square root close to the function $f(X) = X$. See Figure 19.2.

Do not worry if you do not understand why our first and second formulations of Riemann's Hypothesis are equivalent. Our aim, in offering the second formulation – a way of phrasing Riemann's guess that mathematicians know to be equivalent to the first one – is to celebrate the variety of equivalent ways we have to express Riemann's proposed answers to the question "How many

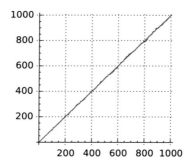

Figure 19.2. The newly constructed staircase is close to the 45 degree line

primes are there?", and to point out that some formulations would reveal a startling simplicity – not immediately apparent – to the behavior of prime numbers, no matter how erratic primes initially appear to us to be. After all, what could be simpler than a 45 degree straight line?

What do Computer Music Files, Data Compression, and Prime Numbers have to do with Each Other?

Sounds of all sorts – and in particular the sounds of music – travel as vibrations of air molecules at roughly 768 miles per hour. These vibrations – fluctuations of pressure – are often represented, or "pictured," by a graph where the horizontal axis corresponds to time, and the vertical axis corresponds to pressure at that time. The very purest of sounds – a single sustained note – would look something like this (called a "sine wave") when pictured (see Figure 20.1), so that if you fixed your position somewhere and measured air pressure due to this sound at that position, the peaks correspond to the times when the varying air pressure is maximal or minimal and the zeroes to the times when it is normal pressure.

You'll notice that there are two features to the graph in Figure 20.1.

1. *The height of the peaks of this sine wave:* This height is referred to as the **amplitude** and corresponds to the *loudness* of the sound.
2. *The number of peaks per second:* This number is referred to as the **frequency** and corresponds to the *pitch* of the sound.

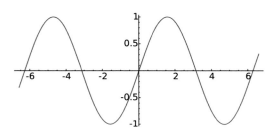

Figure 20.1. Graph of a sine wave

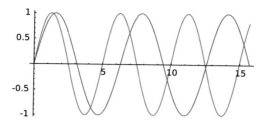

Figure 20.2. Graph of two sine waves with different frequencies

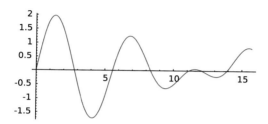

Figure 20.3. Graph of sum of the two sine waves with different frequencies

Of course, music is rarely – perhaps never – just given by a single pure sustained note and nothing else. A next most simple example of a sound would be a simple chord (say a C and an E played together on some electronic instrument that could approximate pure notes). Its graph would be just the *sum* of the graphs of each of the pure notes (see Figures 20.2 and 20.3).

So the picture of the changing frequencies of this chord would be already a pretty complicated configuration. What we have described in these graphs are two sine waves (our C and our E) when they are played *in phase* (meaning they start at the same time) but we could also delay the onset of the E note and play them with some different phase relationship, for example, as illustrated in Figures 20.4 and 20.5.

So, *all you need* to reconstruct the chord graphed above is to know five numbers:

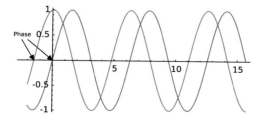

Figure 20.4. Graph of two "sine" waves with different phases

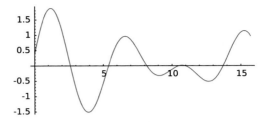

Figure 20.5. Graph of the sum of the two "sine" waves with different frequencies and phases

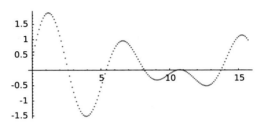

Figure 20.6. Graph of sampling of a sound wave

- the two frequencies – the collection of frequencies that make up the sound is called the *spectrum* of the sound,
- the two *amplitudes* of each of these two frequencies,
- the *phase* between them.

Now suppose you came across such a sound as pictured in Figure 20.5 and wanted to "record it." Well, one way would be to sample the amplitude of the sound at many different times, as for example in Figure 20.6.

Then, fill in the rest of the points to obtain Figure 20.7.

But this sampling would take an enormous amount of storage space, at least compared to storing five numbers, as explained above! Current audio compact discs do their sampling 44,100 times a second to get a reasonable quality of sound.

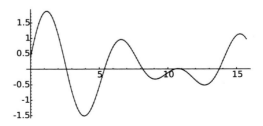

Figure 20.7. Graph obtained from Figure 20.6 by filling in the rest of the points

Figure 20.8. Jean Baptiste Joseph Fourier (1768–1830). Courtesy of the Smithsonian Libraries, Washington, D.C

Another way is to simply record the *five* numbers: the *spectrum, amplitudes,* and *phase.* Surprisingly, this seems to be roughly the way our ear processes such a sound when we hear it.[1]

Even in this simplest of examples (our pure chord: the pure note C played simultaneously with pure note E) the *efficiency of data compression* that is the immediate bonus of analyzing the picture of the chords as built *just* with the five numbers giving *spectrum, amplitudes,* and *phase* is staggering.

This type of analysis, in general, is called *Fourier Analysis* and is one of the glorious chapters of mathematics. One way of picturing *spectrum* and *amplitudes* of a sound is by a bar graph which might be called the *spectral picture* of the sound, the horizontal axis depicting frequency and the vertical one depicting amplitude: the height of a bar at any frequency is proportional to the amplitude of that frequency "in" the sound.

So our CE chord would have the spectral picture in Figure 20.9.

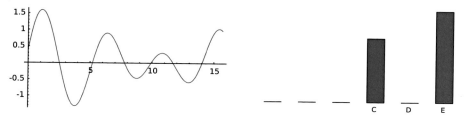

Figure 20.9. Spectral picture of a CE chord

---

[1] We recommend downloading Dave Benson's marvelous book *Music: A Mathematical Offering* from https://homepages.abdn.ac.uk/mth192/pages/html/ maths-music.html. This is free, and gives a beautiful account of the superb mechanism of hearing, and of the mathematics of music.

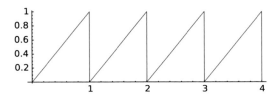

Figure 20.10. Graph of sawtooth wave

This spectral picture ignores the phase but is nevertheless a very good portrait of the sound. The spectral picture of a graph gets us to think of that graph as "built up by the superposition of a bunch of pure waves," and if the graph is complicated enough we may very well need *infinitely* many pure waves to build it up! Fourier analysis is a mathematical theory that allows us to start with any graph – we are thinking here of graphs that picture sounds, but any graph will do – and actually compute its spectral picture (and even keep track of phases).

The operation that starts with a graph and goes to its spectral picture that records the frequencies, amplitudes, and phases of the pure sine waves that, together, compose the graph is called the *Fourier transform* and nowadays there are very fast procedures for getting accurate *Fourier transforms* (meaning accurate spectral pictures including information about phases) by computer [13].

The theory behind this operation (Fourier transform giving us a spectral analysis of a graph) is quite beautiful, but equally impressive is how – given the power of modern computation – you can immediately perform this operation for yourself to get a sense of how different wave-sounds can be constructed from the superposition of pure tones.

The *sawtooth* wave in Figure 20.10 has a spectral picture, its Fourier transform, given in Figure 20.11.

Suppose you have a complicated sound wave, say as in Figure 20.12, and you want to record it. Standard audio CDs record their data by intensive sampling as we mentioned. In contrast, current MP3 audio compression technology uses Fourier transforms plus sophisticated algorithms based on knowledge of which

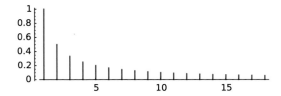

Figure 20.11. The Spectrum of the sawtooth wave has a spike of height $1/k$ at each integer k

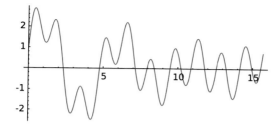

Figure 20.12. A Complicated sound wave

frequencies the human ear can hear. With this, MP3 technology manages to get a compression factor of 8–12 with little *perceived* loss in quality, so that you can fit your entire music collection on your phone, instead of just a few of your favorite CDs.

21 The Word "Spectrum"

It is worth noting how often this word appears in scientific literature, with an array of different uses and meanings. It comes from Latin where its meaning is "image," or "appearance," related to the verb meaning *to look* (the older form being *specere* and the later form *spectare*). In most of its meanings, nowadays, it has to do with some procedure, an analysis, allowing one to see clearly the constituent parts of something to be analyzed, these constituent parts often organized in some continuous scale, such as in the discussion of the previous chapter.

Figure 21.1. Rainbow coming up over a hillside © iStock.com

The Oxford English Dictionary lists, as one of its many uses:

Used to classify something, or suggest that it can be classified, in terms of its position on a scale between two extreme or opposite points.

This works well for the color spectrum, as initiated by Newton (as in the figure above, sunlight is separated by a prism into a rainbow continuum of colors): an analysis of white light into its components. Or in mass spectrometry, where beams of ions are separated (analyzed) according to their mass/charge ratio and the mass spectrum is recorded on a photographic plate or film. Or in the recording of the various component frequencies, with their corresponding intensities of some audible phenomenon.

In mathematics the word has found its use in many different fields, the most basic use occurring in Fourier analysis, which has as its goal the aim of either *analyzing* a function $f(t)$ as being comprised of simpler (specifically: sine and cosine) functions, or *synthesizing* such a function by combining simpler functions to produce it. The understanding here and in the previous chapter, is that an analysis of $f(t)$ as built up of simpler functions is meant to provide a significantly clearer image of the constitution of $f(t)$. If, to take a very particular example, the simpler functions that are needed for the synthesis of $f(t)$ are of the form $a\cos(\theta t)$ (for a some real number which is the *amplitude* or size of the peaks of this periodic function) – and if $f(t)$ is given as the limit of

$$a_1 \cos(\theta_1 t) + a_2 \cos(\theta_2 t) + a_3 \cos(\theta_3 t) \ldots$$

for a sequence of real numbers $\theta_1, \theta_2, \theta_3, \ldots$ (i.e., these simpler functions are functions of *periods* $\frac{2\pi}{\theta_1}, \frac{2\pi}{\theta_2}, \frac{2\pi}{\theta_3}, \ldots$) it is natural to call these θ_i the **spectrum** of $f(t)$. This will eventually show up in our discussion below of trigonometric sums and the *Riemann spectrum*.

22 Spectra and Trigonometric Sums

As we saw in Chapter 20, a pure tone can be represented by a periodic *sine wave* – a function of time $f(t)$ – the equation of which might be:

$$f(t) = a \cdot \cos(b + \theta t).$$

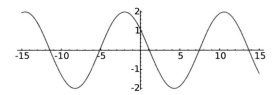

Figure 22.1. Plot of the periodic sine wave $f(t) = 2 \cdot \cos(1 + t/2)$

The angle θ determines the *frequency* of the periodic wave, where the larger θ is the higher the "pitch." The coefficient a determines the envelope of size of the periodic wave, and we call it the *amplitude* of the periodic wave.

Sometimes we encounter functions $F(t)$ that are not pure tones, but that can be expressed as (or we might say "decomposed into") a finite sum of pure tones. For example:

$$F(t) = a_1 \cdot \cos(b_1 + \theta_1 t) + a_2 \cdot \cos(b_2 + \theta_2 t) + a_3 \cdot \cos(b_3 + \theta_3 t)$$

We refer to such functions $F(t)$ as in Figure 22.2 as *finite trigonometric sums*, because – well – they are. In this example, there are three frequencies involved – i.e., $\theta_1, \theta_2, \theta_3$ – and we say that *the spectrum of* $F(t)$ is the set of these

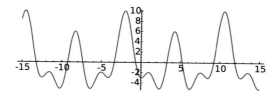

Figure 22.2. Plot of the sum $5\cos(-t-2) + 2\cos(t/2+1) + 3\cos(2t+4)$

frequencies, i.e.,

$$\text{The spectrum of } F(t) = \{\theta_1, \theta_2, \theta_3\}.$$

More generally we might consider a sum of any finite number of pure cosine waves – or in a moment we'll also see some infinite ones as well. Again, for these more general trigonometric sums, their *spectrum* will denote the set of frequencies that compose them.

23 The Spectrum and the Staircase of Primes

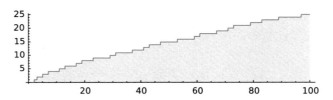

Figure 23.1. The Staircase of primes

In view of the amazing data-compression virtues of Fourier analysis, it isn't unnatural to ask these questions:

- Is there a way of using Fourier analysis to better understand the complicated picture of the staircase of primes?
- Does this staircase of primes (or, perhaps, some tinkered version of the staircase that contains the same basic information) have a *spectrum*?
- If such a *spectrum* exists, can we compute it conveniently, just as we have done for the saw-tooth wave above, or for the major third CE chord?
- Assuming the spectrum exists, and is computable, will our understanding of this spectrum allow us to reproduce all the pertinent information about the placement of primes among all whole numbers, elegantly and faithfully?
- And here is a most important question: will that spectrum show us order and organization lurking within the staircase that we would otherwise be blind to?

Strangely enough, it is towards questions like these that Riemann's Hypothesis takes us. We began with the simple question about primes: how to count them, and are led to ask for profound, and hidden, regularities in structure.

24 To Our Readers of Part I

The statement of the Riemann Hypothesis – admittedly as elusive as before – has, at least, been expressed elegantly and more simply, given our new staircase that approximates (conjecturally with *essential square root accuracy*) a 45 degree straight line.

We have offered two equivalent formulations of the Riemann Hypothesis, both having to do with the manner in which the prime numbers are situated among all whole numbers.

In doing this, we hope that we have convinced you that – in the words of Don Zagier – primes seem to obey no other law than that of chance and yet exhibit stunning regularity. This is the end of Part I of our book, and is largely the end of our main mission, to explain – in elementary terms – *what is Riemann's Hypothesis?*

For readers who have at some point studied differential calculus, in Part II we shall discuss Fourier analysis, a fundamental tool that will be used in Part III where we show how Riemann's hypothesis provides a key to some deeper structure of the prime numbers, and to the nature of the laws that they obey. We will – if not explain – at least hint at how the above series of questions have been answered so far, and how the Riemann Hypothesis offers a surprise for the last question in this series.

PART II

Distributions

How Calculus Manages to Find the Slopes of Graphs That Have No Slopes

Differential calculus, initially the creation of Newton and Leibniz in the 1680s, acquaints us with *slopes* of graphs of functions of a real variable. So, to discuss this we should say a word about what a *function* is, and what its *graph* is.

A **function** (let us refer to it in this discussion as *f*) is often described as a *kind of machine* that for any specific input numerical value *a* will give, as output, a well-defined numerical value.

This "output number" is denoted $f(a)$ and is called *the "value" of the function f at a*. For example, the *machine that adds* 1 *to any number* can be thought of as the function *f* whose value at any *a* is given by the equation $f(a) = a + 1$. Often we choose a letter – say, *X* – to stand for a "general number" and we

Figure 25.1. Isaac Newton and Gottfried Leibniz created Calculus. "Sir Isaac Newton, English mathematician and physicist, 1689. 1863, after the original by Sir Godfrey Kneller of 1689" by Thomas Barlow (1824–1895) © SSPL/Science Museum / Art Resource, NY. "Portrait of Gottfried Wilhelm von Leibniz." c. 1700, Courtesy of the Smithsonian Libraries, Washington, D.C

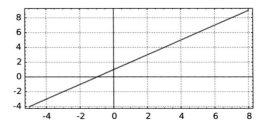

Figure 25.2. Graph of the function $f(a) = a + 1$

denote the function f by the symbol $f(X)$ so that this symbolization allows to "substitute for X any specific number a" to get its value $f(a)$.

The **graph** of a function provides a vivid visual representation of the function in the Euclidean plane where over every point a on the x-axis you plot a point above it of "height" equal to the value of the function at a, i.e., $f(a)$. In Cartesian coordinates, then, you are plotting points $(a, f(a))$ in the plane where a runs through all real numbers.

In this book we will very often be talking about "graphs" when we are also specifically interested in the functions – of which they are the graphs. We will use these words almost synonymously since we like to adopt a very visual attitude towards the behavior of the functions that interest us.

Figure 25.3 illustrates a function (blue), the slope at a point (green straight line), and the derivative (red) of the function; the red derivative is the function whose value at a point is the slope of the blue function at that point. Differential calculus explains to us how to calculate slopes of graphs, and finally, shows us the power that we then have to answer problems we could not answer if we couldn't compute those slopes.

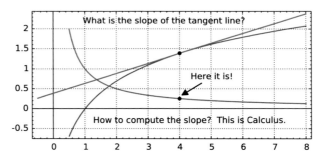

Figure 25.3. Calculus

Usually, in elementary calculus classes we are called upon to compute slopes only of smooth graphs, i.e., graphs that actually *have* slopes at each of their points, such as in the illustration just above. What could calculus possibly do if confronted with a graph that has *jumps*, such as in Figure 25.4:

$$f(x) = \begin{cases} 1 & x \le 3 \\ 2 & x > 3. \end{cases}$$

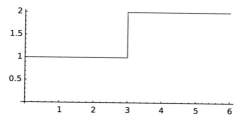

Figure 25.4. The graph of the function $f(x)$ above that jumps – it is 1 up to 3 and then 2 after that point

(Note that for purely aesthetic reasons, we draw a vertical line at the point where the jump occurs, though technically that vertical line is not part of the graph of the function.)

The most comfortable way to deal with the graph of such a function is to just approximate it by a differentiable function as in Figure 25.5.

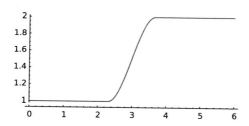

Figure 25.5. A picture of a smooth graph approximating the graph that is 1 up to some point x and then 2 after that point, the smooth graph being flat mostly

Then take the *derivative* of that smooth function. Of course, this is just an approximation, so we might try to make a better approximation, which we do in each successive graph starting with Figure 25.6 below.

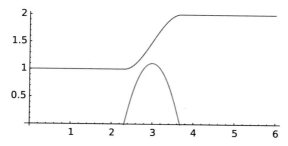

Figure 25.6. A picture of the derivative of a smooth approximation to a function that jumps

Note that – as you would expect – in the range where the initial function is constant, its derivative is zero. In the subsequent figures, our initial function will be *nonconstant* for smaller and smaller intervals about the origin. Note also

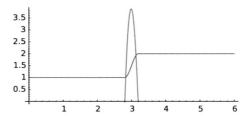

Figure 25.7. Second picture of the derivative of a smooth approximation to a function that jumps

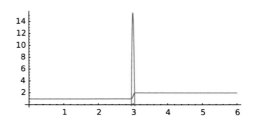

Figure 25.8. Third picture of the derivative of a smooth approximation to a function that jumps

that, in our series of pictures below, we will be successively rescaling the y-axis; all our initial functions have the value 1 for "large" negative numbers and the value 2 for large positive numbers.

Notice what is happening: as the approximation gets better and better, the derivative will be zero mostly, with a blip at the point of discontinuity, and the blip will get higher and higher. In each of these pictures, for any interval of real numbers $[a, b]$ the total area under the red graph over that interval is equal to

the height of the blue graph at $x = b$
minus
the height of the blue graph at $x = a$.

This is a manifestation of one of the fundamental facts of life of calculus relating a function to its derivative:

Given any real-valued function $F(x)$ – that has a derivative – for any interval of real numbers $[a, b]$ the total signed[1] area between the graph and the horizontal axis of the derivative of $F(x)$ over that interval is equal to $F(b) - F(a)$.

What happens if we take the series of figures 25.6–25.9, etc., *to the limit?* This is quite curious:

---

[1] When $F(x) < 0$ we count that area as negative.

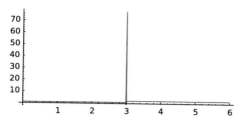

Figure 25.9. Fourth picture of the derivative of a smooth approximation to a function that jumps

- **the series of red graphs:** these are getting thinner and thinner and higher and higher: can we make any sense of what the red graph might mean in the limit (even though the only picture of it that we have at present makes it infinitely thin and infinitely high)?
- **the series of blue graphs:** these are happily looking more and more like the tame Figure 25.4.

Each of our red graphs is the derivative of the corresponding blue graph. It is tempting to think of the limit of the red graphs – whatever we might construe this to be – as standing for the derivative of the limit of the blue graphs, i.e., of the graph in Figure 25.4.

Well, the temptation is so great that, in fact, mathematicians and physicists of the early twentieth century struggled to give a meaning to things like *the limit of the red graphs* – such things were initially called **generalized functions** which might be considered the derivative of *the limit of the blue graphs*, i.e., of the graph of Figure 25.4.

Of course, to achieve progress in mathematics, all the concepts that play a role in the theory have to be unambiguously defined, and it took a while before *generalized functions* such as the limit of our series of red graphs had been rigorously introduced.

But many of the great moments in the development of mathematics occur when mathematicians – requiring some concept not yet formalized – work with the concept tentatively, dismissing – if need be – mental tortures, in hopes that the experience they acquire by working with the concept will eventually help to put that concept on sure footing. For example, early mathematicians (Newton, Leibniz) – in replacing approximate speeds by instantaneous velocities by passing to limits – had to wait a while before later mathematicians (e.g., Weierstrass) gave a rigorous foundation for what they were doing.

Karl Weierstrass, who worked during the latter part of the nineteenth century, was known as the "father of modern analysis." He oversaw one of the glorious moments of rigorization of concepts that were long in use, but never before systematically organized. He, and other analysts of the time, were interested in providing a rigorous language to talk about *functions* and more specifically *continuous functions* and *smooth* (i.e., *differentiable*) functions. They wished

Figure 25.10. Karl Weierstrass (1815–1897) Courtesy of the Smithsonian Libraries, Washington, D.C. Laurent Schwartz (1915–2002) Author: Konrad Jacobs, *Source: Archives of the Mathematisches Forschungsinstitut Oberwolfach.*

to have a firm understanding of limits (i.e., of sequences of numbers, or of functions).

For Weierstrass and his companions, even though the functions they worked with needn't be smooth, or continuous, at the very least, the functions they studied had *well-defined – and usually finite – values.* But our "limit of red graphs" is not so easily formalized as the concepts that occupied the efforts of Weierstrass.

Happily however, this general process of approximating discontinuous functions more and more exactly by smooth functions, and taking their derivatives to get the blip-functions as we have just seen in the red graphs above was eventually given a mathematically rigorous foundation; notably, by the French mathematician, Laurent Schwartz who provided a beautiful theory that we will not go into here, that made perfect sense of "generalized functions" such as our limit of the series of red graphs, and that allows mathematicians to work with these concepts with ease. These "generalized functions" are called *distributions* in Schwartz's theory [14].

26 Distributions: Sharpening our Approximating Functions even if we have to Let them Shoot Out to Infinity

The curious *limit of the red graphs* of the previous section, which you might be tempted to think of as a "blip-function" that vanishes for t nonzero and is somehow "infinite" (whatever that means) at 0 is an example of a *generalized function* (in the sense of the earlier mathematicians) or a *distribution* in the sense of Laurent Schwartz.

This particular *limit of the red graphs* also goes by another name (it is officially called a Dirac δ-function (see [15]), the adjective "Dirac" being in honor of the physicist who first worked with this concept, the "δ" being the symbol he assigned to these objects). The noun "function" should be in quotation marks for, properly speaking, the Dirac δ-function is not – as we have explained above – a bona fide function but rather a distribution.

Now may be a good time to summarize what the major difference is between *honest functions* and *generalized functions* or *distributions*.

Figure 26.1. Paul Adrien Maurice Dirac (1902–1984) © Peter Lofts Photography / National Portrait Gallery, London

$$\int_a^b f(t)dt.$$

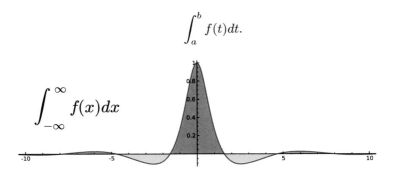

$$\int_{-\infty}^{\infty} f(x)dx$$

Figure 26.2. This figure illustrates $\int_{-\infty}^{\infty} f(x)dx$, which is the signed area between the graph of $f(x)$ and the x-axis, where area below the x-axis (yellow) counts negative, and area above (grey) is positive

An honest (by which we mean *integrable*) function of a real variable $f(t)$ possesses two "features."

- **It has values.** That is, at any real number t, e.g., $t = 2$ or $t = 0$ or $t = \pi$ etc., our function has a definite real number value ($f(2)$ or $f(0)$ or $f(\pi)$ etc.) *and if we know all those values we know the function.*
- **It has areas under its graph.** If we are given any interval of real numbers, say the interval between a and b, we can talk unambiguously about the area "under" the graph of the function $f(t)$ over the interval between a and b. That is, in the terminology of integral calculus, we can talk of *the integral of $f(t)$ from a to b.* And in the notation of calculus, this – thanks to Leibniz – is elegantly denoted

$$\int_a^b f(t)dt.$$

In contrast, a *generalized function* or *distribution*

- **may not have "definite values"** at all real numbers if it is not an honest function. Nevertheless,
- **It has well-defined areas under portions of its "graph."** If we are given any interval of real numbers, say the (open) interval between a and b, we can still talk unambiguously about the *area "under" the graph of the generalized function $D(t)$ over the interval between a and b* and we will denote this – extending what we do in ordinary calculus – by the symbol

$$\int_a^b D(t)dt.$$

This description is important to bear in mind and it gives us a handy way of thinking about "generalized functions" (i.e., distributions) as opposed to functions: when we consider an (integrable) function of a real variable, $f(t)$, we may invoke its *value* at every real number and for every interval $[a, b]$ we may consider the quantity $\int_a^b f(t)dt$. BUT when we are given a generalized function $D(t)$ we *only* have at our disposal the latter quantities. In fact, a generalized function of a real variable $D(t)$ is (formally) nothing more than a *rule* that assigns to any finite interval $(a, b]$ $(a \leq b)$ a quantity that we might denote $\int_a^b D(t)dt$ and that *behaves as if it were the integral of a function* and in particular – for three real numbers $a \leq b \leq c$ we have the additivity relation

$$\int_a^c D(t)dt = \int_a^b D(t)dt + \int_b^c D(t)dt.$$

SO, any honest function integrable over finite intervals clearly *is* a distribution (forget about its values!) but . . . there are many more generalized functions, and including them in our sights gives us a very important tool.

It is natural to talk, as well, of Cauchy sequences, and limits, of distributions. We'll say that such a sequence $D_1(t), D_2(t), D_3(t), \ldots$ is a **Cauchy sequence** if for every interval $[a, b]$ the quantities

$$\int_a^b D_1(t)dt, \quad \int_a^b D_2(t)dt, \quad \int_a^b D_3(t)dt, \ldots$$

form a Cauchy sequence of real numbers (so for any $\varepsilon > 0$ eventually all terms in the sequence of real numbers are within ε of each other). Now, any Cauchy sequence of distributions *converges to a limiting distribution $D(t)$* which is defined by the rule that for every interval $[a, b]$,

$$\int_a^b D(t)dt = \lim_{i \to \infty} \int_a^b D_i(t)dt.$$

If, by the way, you have an infinite sequence – say – of honest, continuous, functions that converges uniformly to a limit (which will again be a continuous function) then that sequence certainly converges – in the above sense – to the same limit when these functions are viewed as generalized functions. BUT, there are many important occasions where your sequence of honest continuous functions doesn't have that convergence property and *yet* when these continuous functions are viewed as generalized functions they do converge to some generalized function as a limit. We will see this soon when we get back to the "sequence of the red graphs." This sequence **does** converge (in the above sense) to the Dirac δ-function when these red graphs are thought of as a sequence of generalized functions.

The integral notation for distribution is very useful, and allows us the flexibility to define, for nice enough – and honest – functions $c(t)$ useful expressions such as

$$\int_a^b c(t)D(t).$$

For example, the Dirac δ-function we have been discussing (i.e., the limit of the red graphs of Chapter 25) *is* an honest function away from $t = 3$ and – in fact – is the "trivial function" zero away from 3. And at 3, we may *say* that it has the "value" infinity, in honor of it being the limit of blip functions getting taller and

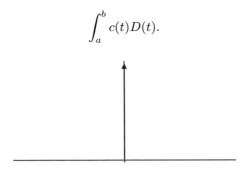

$$\int_a^b c(t)D(t).$$

Figure 26.3. The Dirac δ-"function" (actually distribution), where we draw a vertical arrow to illustrate the delta function with support at a given point

taller at 3. The feature that pins it down as a distribution is given by its behavior relative to the second feature above, the area of its graph over the open interval (a, b) between a and b.

- If 3 is not in the open interval spanned by a and b, then the "area under the graph of our Dirac δ-function" over the interval (a, b) is 0.
- If 3 is in the open interval (a, b), then the "area under the graph of our Dirac δ-function" is 1 – in notation

$$\int_a^b \delta = 1.$$

We sometimes summarize the fact that these areas vanish so long as 3 is not included in the interval we are considering by saying that the **support** of this δ-function is "at 3."

Once you're happy with *this* Dirac δ-function, you'll also be happy with a Dirac δ-function – call it δ_x – with support concentrated at any specific real number x. This δ_x vanishes for $t \neq x$ and intuitively speaking, has an *infinite blip* at $t = x$.

So, the original delta-function we were discussing, i.e., $\delta(t)$ would be denoted $\delta_3(t)$.

A question: If you've never seen distributions before, but know the Riemann integral, can you guess at what the definition of $\int_a^b c(t)D(t)$ is, and can you formulate hypotheses on $c(t)$ that would allow you to endow this expression with a definite meaning?

A second question: If you have not seen distributions before, and have answered the first question above, let $c(t)$ be an honest function for which your definition of

$$\int_a^b c(t)D(t)$$

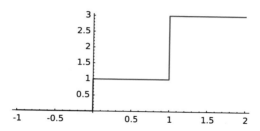

Figure 26.4. The staircase function that is 0 for $t \leq 0$, 1 for $0 < t \leq 1$ and 3 for $1 < t \leq 2$ has derivative $\delta_0 + 2\delta_1$

applies. Now let x be a real number. Can you use your definition to compute

$$\int_{-\infty}^{+\infty} c(t)\delta_x(t)?$$

The answer to this second question, by the way, is: $\int_{-\infty}^{+\infty} c(t)\delta_x(t) = c(x)$. This will be useful in the later sections!

The theory of distributions gives a partial answer to the following funny question:

How in the world can you "take the derivative" of a function $F(t)$ that doesn't have a derivative?

The short answer to this question is that *this derivative $F'(t)$ which doesn't exist as a function may exist as a distribution*. What then is the integral of that distribution? Well, it is given by the original function!

$$\int_a^b F'(t)\,dt = F(b) - F(a).$$

Let us practice this with simple staircase functions. For example, what is the *derivative* – in the sense of the theory of distributions – of the function in Figure 26.4? **Answer.** $\delta_0 + 2\delta_1$.

We'll be dealing with much more complicated staircase functions in the next chapter, but the general principles discussed here will nicely apply there [16].

27 Fourier Transforms: Second Visit

In Chapter 20 above we wrote:

> The operation that starts with a graph and goes to its spectral picture that records the frequencies, amplitudes, and phases of the pure sine waves that, together, compose the graph is called the **Fourier transform**.

Now let's take a closer look at this operation *Fourier transform*.

We will focus our discussion on an **even** function $f(t)$ of a real variable t. "**Even**" means that its graph is symmetric about the y-axis; that is, $f(-t) = f(t)$. See Figure 27.1.

When we get to apply this discussion to the *staircase of primes* $\pi(t)$ or the *tinkered staircase of primes* $\psi(t)$, both of which being defined only for positive values of t, then we would "lose little information" in our quest to understand them if we simply "symmetrized their graphs" by defining their values on negative numbers $-t$ via the formulas $\pi(-t) = \pi(t)$ and $\psi(-t) = \psi(t)$ thereby turning each of them into *even functions*.

The idea behind the Fourier transform is to express $f(t)$ as *made up out of sine and cosine wave functions*. Since we have agreed to consider only even

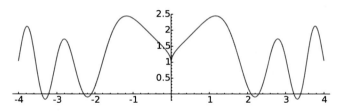

Figure 27.1. The graph of an even function is symmetrical about the y-axis

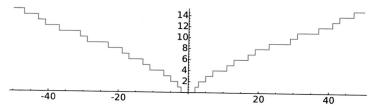

Figure 27.2. Even extension of the staircase of primes

functions, we can dispense with the sine waves – they won't appear in our Fourier analysis – and ask how to reconstruct $f(t)$ as a *sum* (with coefficients) of cosine functions (if only finitely many frequencies occur in the spectrum of our function) or more generally, as an *integral* if the spectrum is more elaborate. For this work, we need a little machine that tells us, for each real number θ, whether or not θ is in the spectrum of $f(t)$, and if so, what the amplitude is of the cosine function $\cos(\theta t)$ that occurs in the Fourier expansion of $f(t)$ – this amplitude answers the awkwardly phrased question:

How much $\cos(\theta t)$ *"occurs in"* $f(t)$?

We will denote this amplitude by $\hat{f}(\theta)$, and refer to it as **the Fourier transform** of $f(t)$. The **spectrum**, then, of $f(t)$ is the set of all frequencies θ where the amplitude is nonzero.

Now in certain easy circumstances – specifically, if $\int_{-\infty}^{+\infty} |f(t)|dt$ (exists, and) is finite – integral calculus provides us with an easy construction of that machine (see Figure 27.3); namely:

$$\hat{f}(\theta) = \int_{-\infty}^{+\infty} f(t)\cos(-\theta t)dt.$$

This concise machine manages to "pick out" just the part of $f(t)$ that has frequency θ! It provides for us the *analysis* part of the Fourier analysis of our function $f(t)$.

But there is a *synthesis* part to our work as well, for we can reconstruct $f(t)$ from its Fourier transform, by a process intriguingly similar to the

$f(t)\longrightarrow$ $\longrightarrow\hat{f}(\theta)$

Figure 27.3. The Fourier transform machine, which transforms $f(t)$ into $\hat{f}(\theta)$

analysis part; namely: if $\int_{-\infty}^{+\infty} |\hat{f}(\theta)| d\theta$ (exists, and) is finite, we retrieve $f(t)$ by the integral

$$f(t) = \frac{1}{2\pi} \int_{-\infty}^{+\infty} \hat{f}(\theta) \cos(\theta t) d\theta.$$

We are not so lucky to have $\int_{-\infty}^{+\infty} |f(t)| dt$ finite when we try our hand at a Fourier analysis of the staircase of primes, but we'll work around this!

28 What is the Fourier Transform of a Delta Function?

Consider the δ-function that we denoted $\delta(t)$ (or $\delta_0(t)$). This is also the "generalized function" that we thought of as the "limit of the red graphs" in Chapter 26 above. Even though $\delta(t)$ is a distribution and *not* a bona fide function, it is symmetric about the origin, and also

$$\int_{-\infty}^{+\infty} |\delta(t)| dt$$

exists, and is finite (its value is, in fact, 1). All this means that, appropriately understood, the discussion of the previous section applies, and we can *feed* this delta-function into our Fourier Transform Machine (Figure 27.3) to see what frequencies and amplitudes arise in our attempt to express – whatever this means! – the delta-function as a sum, or an integral, of cosine functions.

So what is the Fourier transform, $\hat{\delta}_0(\theta)$, of the delta-function?

Well, the general formula would give us:

$$\hat{\delta}_0(\theta) = \int_{-\infty}^{+\infty} \cos(-\theta t)\delta_0(t)dt$$

and as we mentioned in section 18, for any nice function $c(t)$ we have that the integral of the product of $c(t)$ by the distribution $\delta_x(t)$ is given by the *value* of the function $c(t)$ at $t = x$. So:

$$\hat{\delta}_0(\theta) = \int_{-\infty}^{+\infty} \cos(-\theta t)\delta_0(t)dt = \cos(0) = 1.$$

In other words, the Fourier transform of $\delta_0(t)$ is the constant function

$$\hat{\delta}_0(\theta) = 1.$$

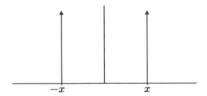

Figure 28.1. The sum $(\delta_x(t) + \delta_{-x}(t))/2$, where we draw vertical arrows to illustrate the Dirac delta functions

One can think of this colloquially as saying that the delta-function is a perfect example of *white noise* in that *every* frequency occurs in its Fourier analysis and they all occur in equal amounts.

To generalize this computation, let us consider for any real number x the symmetrized delta-function with support at x and $-x$, given by

$$d_x(t) = (\delta_x(t) + \delta_{-x}(t))/2$$

in Figure 28.1.

What is the Fourier transform of this $d_x(t)$? The answer is given by making the same computation as we've just made:

$$\hat{d}_x(\theta) = \frac{1}{2}\left(\int_{-\infty}^{+\infty} \cos(-\theta t)\delta_x(t)\,dt + \int_{-\infty}^{+\infty} \cos(-\theta t)\delta_{-x}(t)\,dt\right)$$

$$= \frac{1}{2}\left(\cos(-\theta x) + \cos(+\theta x)\right)$$

$$= \cos(x\theta)$$

To summarize this in ridiculous (!) colloquial terms: *for any frequency θ the amount of $\cos(\theta t)$ you need to build up the generalized function $(\delta_x(t) + \delta_{-x}(t))/2$ is $\cos(x\theta)$.*

So far, so good, but remember that the theory of the Fourier transform has – like much of mathematics – two parts: an *analysis part* and a *synthesis* part. We've just performed the *analysis* part of the theory for these symmetrized delta functions $(\delta_x(t) + \delta_{-x}(t))/2$.

Can we synthesize them – i.e., build them up again – from their Fourier transforms?

We'll leave this, at least for now, as a question for you.

29 Trigonometric Series

Given our interest in the ideas of Fourier, it is not surprising that we'll want to deal with things like

$$F(\theta) = \sum_{k=1}^{\infty} a_k \cos(s_k \cdot \theta)$$

where the s_k are real numbers tending (strictly monotonically) to infinity. These we'll just call **trigonometric series** without asking whether they converge in any sense for all values of θ, or even for *any* value of θ. The s_k's that occur in such a trigonometric series we will call the **spectral values** or for short, the **spectrum** of the series, and the a_k's the (corresponding) **amplitudes**. We repeat that we impose no convergence requirements at all. But we also think of these things as providing "cutoff" finite trigonometric sums, which we think of as functions of two variables, θ and C (the "cutoff") where

$$F(\theta, C) := \sum_{s_k \leq C} a_k \cos(s_k \cdot \theta).$$

These functions $F(\theta, C)$ are finite trigonometric series and therefore "honest functions" having finite values everywhere.

Recall, as in Chapter 28, that for any real number x, we considered the symmetrized delta-function with support at x and $-x$, given by

$$d_x(t) = (\delta_x(t) + \delta_{-x}(t))/2,$$

and noted that the Fourier transform of this $d_x(t)$ is

$$\hat{d}_x(\theta) = \cos(x\theta).$$

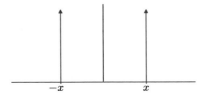

Figure 29.1. The sum $(\delta_x(t) + \delta_{-x}(t))/2$, where we draw vertical arrows to illustrate the Dirac delta functions

It follows, of course, that a cutoff finite trigonometric series, $F(\theta, C)$ associated to an infinite trigonometric series

$$F(\theta) = \sum_{k=1}^{\infty} a_k \cos(s_k \cdot \theta)$$

is the Fourier transform of the distribution

$$D(t, C) := \sum_{s_k \leq C} a_k d_{s_k}(t).$$

Given the discreteness of the set of spectral values s_k ($k = 1, 2, \ldots$) we can consider the infinite sum

$$D(t) := \sum_{k=1}^{\infty} a_k d_{s_k}(t),$$

viewed as distribution playing the role of the "inverse Fourier transform" of our trigonometric series $F(t)$.

Definition 29.1

Say that a trigonometric series $F(\theta)$ has a **spike** at the real number $\theta = \tau \in \mathbf{R}$ if the set of absolute values $|F(\tau, C)|$ as C ranges through positive number cutoffs is unbounded. A real number $\tau \in \mathbf{R}$ is, in contrast, a **non-spike** if those values admit a finite upper bound.

In the chapters that follow we will be exhibiting graphs of trigonometric functions, cutoff at various values of C, that (in our opinion) strongly hint that as C goes to infinity, one gets convergence to certain (discrete) sequences of very interesting *spike values*. To be sure, no finite computation, exhibited by a graph, can *prove* that this is in fact the case. But on the one hand, the vividness of those spikes is in itself worth experiencing, and on the other hand, given RH, there is justification that the *strong hints* are not misleading; for some theoretical background see the endnotes.

30 A Sneak Preview of Part III

In this chapter, as a striking illustration of the type of phenomena that will be studied in Part III, we will consider two infinite trigonometric sums – that seem to be related one to the other in that the *frequencies* of the terms in the one trigonometric sum give the *spike values* of the other, and vice versa: the *frequencies* of the other give the *spike values* of the one: a kind of duality as in the theory of Fourier transforms. We show this duality by exhibiting the graphs of more and more accurate finite approximations (cutoffs) of these infinite sums. More specifically,

1. The first infinite trigonometric sum $F(t)$ is a sum[1] of pure cosine waves with frequencies given by *logarithms of powers of primes* and with amplitudes given by the formula

$$F(t) := -\sum_{p^n} \frac{\log(p)}{p^{n/2}} \cos(t \log(p^n))$$

the summation being over all powers p^n of all prime numbers p.

The graphs of longer and longer finite truncations of these trigonometric sums, as you will see, have "higher and higher peaks" concentrated more and more accurately at a *certain infinite discrete set of real numbers* that we will be referring to as **the Riemann spectrum**, indicated in our pictures below (Figures 30.2–30.5) by the series of vertical red lines.

---

[1] Here we make use of the Greek symbol $\sum$ as a shorthand way of expressing a sum of many terms. We are not requesting this sum to converge.

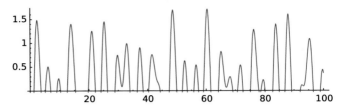

Figure 30.1. Plot of $f(t)$

2. In contrast, the second infinite trigonometric sum $H(t)$ is a sum of pure cosine waves with frequencies given by what we have dubbed above *the Riemann spectrum* and with amplitudes all equal to 1.

$$H(s) := 1 + \sum_{\theta} \cos(\theta \log(s)).$$

These graphs will have "higher and higher peaks" concentrated more and more accurately at **the logarithms of powers of primes** indicated in our pictures below (see Figure 30.6) by the series of vertical blue spikes.

That the series of *blue lines* (i.e., the logarithms of powers of primes) in our pictures below determines – via the trigonometric sums we describe – the series of *red lines* (i.e., what we are calling the spectrum) and conversely is a consequence of the Riemann Hypothesis.

1. Viewing the Riemann spectrum as the spike values of a trigonometric series with frequencies equal to (logs of) powers of the primes:
To get warmed up, let's plot the positive values of the following sum of (co-)sine waves:

$$f(t) = -\frac{\log(2)}{2^{1/2}} \cos(t \log(2)) - \frac{\log(3)}{3^{1/2}} \cos(t \log(3))$$
$$-\frac{\log(2)}{4^{1/2}} \cos(t \log(4)) - \frac{\log(5)}{5^{1/2}} \cos(t \log(5))$$

Look at the peaks of this graph. There is nothing very impressive about them, you might think; but wait, for $f(t)$ is just a very "early" piece of the infinite trigonometric sum $F(t)$ described above.

Let us truncate the infinite sum $F(t)$ taking only finitely many terms, by choosing various "cutoff values" C and forming the finite sums

$$F_{\leq C}(t) := -\sum_{p^n \leq C} \frac{\log(p)}{p^{n/2}} \cos(t \log(p^n))$$

and plotting their positive values. Figures 30.2–30.5 show what we get for a few values of C.

In each of the graphs, we have indicated by red vertical arrows the real numbers that give the values of the *Riemann spectrum* that we will be discussing. These numbers at the red vertical arrows in Figures 30.2–30.5,

$$\theta_1, \theta_2, \theta_3, \ldots$$

are *spike values* – as described in Chapter 29 – of the infinite trigonometric series

$$-\sum_{p^n < C} \frac{\log(p)}{p^{n/2}} \cos(t \log(p^n)).$$

They constitute what we are calling the Riemann spectrum and are key to the staircase of primes [17].

- **The sum with $p^n \leq C = 5$**

  In Figure 30.2 we plot the function $f(t)$ displayed above; it consists of the sum of the first four terms of our infinite sum, and doesn't yet show very much "structure":

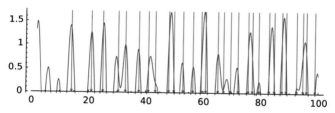

Figure 30.2. Plot of $-\sum_{p^n \leq 5} \frac{\log(p)}{p^{n/2}} \cos(t \log(p^n))$ with arrows pointing to the spectrum of the primes

- **The sum with $p^n \leq C = 20$**

  Something, (don't you agree?) is already beginning to happen in the graph in Figure 30.3:

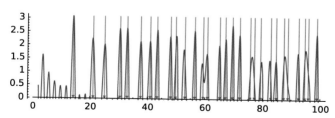

Figure 30.3. Plot of $-\sum_{p^n \leq 20} \frac{\log(p)}{p^{n/2}} \cos(t \log(p^n))$ with arrows pointing to the spectrum of the primes

- **The sum with $p^n \leq C = 50$**

  Note that the high peaks in Figure 30.4 seem to be lining up more accurately with the vertical red lines. Note also that the y-axis has been rescaled.

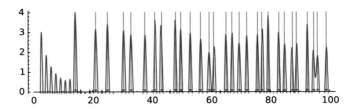

Figure 30.4. Plot of $-\sum_{p^n \leq 50} \frac{\log(p)}{p^{n/2}} \cos(t \log(p^n))$ with arrows pointing to the spectrum of the primes

- **The sum with $p^n \leq C = 500$**
  Here, the peaks are even sharper, and note that again they are higher; that is, we have rescaled the y-axis.

We will pay attention to:

- how the spikes "play out" as we take the sums of longer and longer pieces of the infinite sum of cosine waves above, given by larger and larger cutoffs C,
- how this spectrum of red lines more closely matches the high peaks of the graphs of the positive values of these finite sums,
- how these peaks are climbing higher and higher,
- what relationship these peaks have to the Fourier analysis of the staircase of primes,
- and, equally importantly, what these mysterious red lines signify.

2. **Towards (logs of) powers of the primes, starting from the Riemann spectrum:**

   Here we will be making use of the series of numbers

   $$\theta_1, \theta_2, \theta_3, \ldots$$

   comprising what we called the *spectrum*. We consider the infinite trigonometric series

   $$G(t) := 1 + \cos(\theta_1 t) + \cos(\theta_2 t) + \cos(\theta_3 t) + \cdots$$

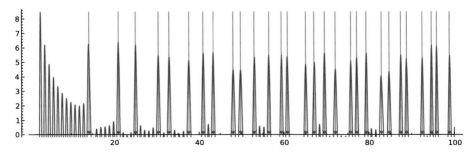

Figure 30.5. Plot of $-\sum_{p^n \leq 500} \frac{\log(p)}{p^{n/2}} \cos(t \log(p^n))$ with arrows pointing to the spectrum of the primes

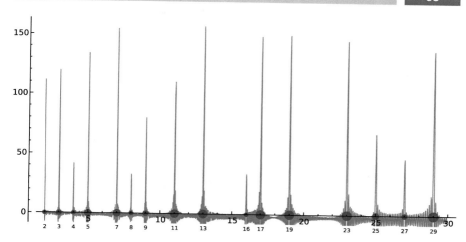

Figure 30.6. Illustration of $- \sum_{i=1}^{1000} \cos(\log(s)\theta_i)$, where $\theta_1 \sim 14.13, \ldots$ are the first 1000 contributions to the spectrum. The red dots are at the prime powers p^n, whose size is proportional to $\log(p)$

or, using the $\sum$ notation,

$$G(t) := 1 + \sum_{\theta} \cos(\theta t)$$

where the summation is over the spectrum, $\theta = \theta_1, \theta_2, \theta_3, \ldots$. Again we will consider finite cutoffs C of this infinite trigonometric sum (on a logarithmic scale),

$$H_{\leq C}(s) := 1 + \sum_{i \leq C} \cos(\log(s)\theta_i)$$

and to see the spikes in $H_{\leq 1000}(s)$ consider Figure 30.6.

This passage – thanks to the Riemann Hypothesis – from spectrum to prime powers and back again via consideration of the "high peaks" in the graphs of the appropriate trigonometric sums provides a kind of visual duality emphasizing, for us, that the information inherent in the wild spacing of prime powers, is somehow "packaged" in the Riemann spectrum, and reciprocally, the information given in that series of mysterious numbers is obtainable from the sequence of prime powers.

The Riemann Spectrum of the Prime Numbers

31 On Losing No Information

To manage to repackage the "same" data in various ways – where each way brings out some features that would be kept in the shadows if the data were packaged in some different way – is a high art, in mathematics. In a sense *every* mathematical equation does this, for the "equal sign" in the middle of the equation tells us that even though the two sides of the equation may seem different, or have different shapes, they are nonetheless "the same data." For example, the equation

$$\log(XY) = \log(X) + \log(Y)$$

which we encountered earlier in Chapter 10, is just two ways of looking at the same thing, yet it was the basis for much manual calculation for several centuries.

Now, the problem we have been concentrating on, in this book, has been – in effect – to understand the pattern, if we can call it that, given by the placement of prime numbers among the natural line-up of all whole numbers.

There are, of course, many ways for us to present this basic pattern. Our initial strategy was to focus attention on the *staircase of primes* which gives us a vivid portrait, if you wish, of the order of appearance of primes among all numbers.

As we have already hinted in the previous sections, however, there are various ways open to us to tinker with – and significantly modify – our staircase *without losing the essential information it contains.* Of course, there is always

Figure 31.1. Prime numbers up to 37

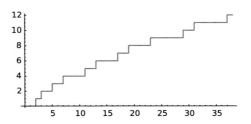

Figure 31.2. Prime numbers up to 37

the danger of modifying things in such a way that "retrieval" of the original data becomes difficult. Moreover, we had better remember every change we have made if we are to have any hope of retrieving the original data!

With this in mind, let us go back to Chapter 18 (discussing the staircase of primes) and Chapter 19, where we tinkered with the original staircase of primes – alias: the graph of $\pi(X)$ – to get $\psi(X)$ whose risers look – from afar – as if they approximated the 45 degree staircase.

At this point we'll do some further carpentry on $\psi(X)$ *without destroying the valuable information it contains*. We will be replacing $\psi(X)$ by a generalized function, i.e., a distribution, which we denote $\Phi(t)$ that has support at all positive integral multiples of logs of prime numbers, and is zero on the complement of that discrete set. Recall that by definition, a discrete subset S of real numbers is the **support** of a function, or of a distribution, if the function vanishes on the complement of S and doesn't vanish on the complement of any proper subset of S.

Given the mission of our book, it may be less important for us to elaborate on the construction of $\Phi(t)$ than it is **(a)** to note that $\Phi(t)$ contains all the valuable information that $\psi(X)$ has and **(b)** to pay close attention to the spike values of the trigonometric series that is the Fourier transform of $\Phi(t)$.

For the definition of the distribution $\Phi(t)$ see the end-note [18].

A distribution that has a discrete set of real numbers as its support – as $\Phi(t)$ does – we sometimes like to call **spike distributions** since the pictures of functions approximating it tend to look like a series of spikes.

We have then before us a spike distribution with support at integral multiples of logarithms of prime numbers, and this generalized function retains the essential information about the placement of prime numbers among all whole numbers, and will be playing a major role in our story: knowledge of the placement of the "blips" constituting this distribution (its support), being at integral multiples of logs of prime numbers, would allow us to reconstruct the position of the prime numbers among all numbers. Of course there are many other ways to package this vital information, so we must explain our motivation for subjecting our poor initial staircase to the particular series of brutal acts of distortion that we described, which ends up with the distribution $\Phi(t)$.

32 Going from the Primes to the Riemann Spectrum

We discussed the nature of the Fourier transform of (symmetrized) δ-functions in Chapter 28. In particular, recall the "spike function"

$$d_x(t) = (\delta_x(t) + \delta_{-x}(t))/2$$

that has support at the points x and $-x$. We mentioned that its Fourier transform, $\hat{d}_x(\theta)$, is equal to $\cos(x\theta)$ (and gave some hints about why this may be true).

Our next goal is to work with the much more interesting "spike function"

$$\Phi(t) = e^{-t/2}\Psi'(t),$$

which was one of the generalized functions that we engineered in Chapter 31, and that has support at all nonnegative integral multiples of logarithms of prime numbers.

As with any function – or generalized function – defined for non-negative values of t, we can "symmetrize it" (about the t-axis) which means that we can define it on negative real numbers by the equation

$$\Phi(-t) = \Phi(t).$$

Let us make that convention, thereby turning $\Phi(t)$ into an *even* generalized function, as illustrated in Figure 32.1. (An **even** function on the real line is a function that takes the same value on any real number and its negative as in the formula above.)

We may want to think of $\Phi(t)$ as a limit of this sequence of distributions,

$$\Phi(t) = \lim_{C \to \infty} \Phi_{\leq C}(t)$$

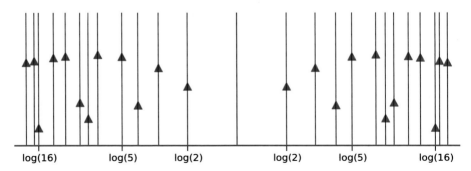

log(16)   log(5)   log(2)   log(2)   log(5)   log(16)

Figure 32.1. $\Phi(t)$ is a sum of Dirac delta functions at the logarithms of prime powers p^n weighted by $p^{-n/2}\log(p)$ (and $\log(2\pi)$ at 0)

where $\Phi_{\leq C}(t)$ is the following finite linear combination of (symmetrized) δ-functions $d_x(t)$:

$$\Phi_{\leq C}(t) := 2 \sum_{\text{prime powers } p^n \leq C} p^{-n/2}\log(p)d_{n\log(p)}(t).$$

Since the Fourier transform of $d_x(t)$ is $\cos(x\theta)$, the Fourier transform of each $d_{n\log(p)}(t)$ is $\cos(n\log(p)\theta)$, so the Fourier transform of $\Phi_{\leq C}(t)$ is

$$\hat\Phi_{\leq C}(\theta) := 2 \sum_{\text{prime powers } p^n \leq C} p^{-n/2}\log(p)\cos(n\log(p)\theta).$$

So, following the discussion in Chapter 29 above, we are dealing with the cutoffs at finite points C of the *trigonometric series*[1]

$$\hat\Phi(\theta) := 2 \sum_{\text{prime powers } p^n} p^{-n/2}\log(p)\cos(n\log(p)\theta).$$

For example, when $C = 3$, we have the rather severe cutoff of these trigonometric series: $\hat\Phi_{\leq 3}(\theta)$ takes account *only* of the primes $p = 2$ and $p = 3$:

$$\hat\Phi_{\leq 3}(\theta) = \frac{2}{\sqrt{2}}\log(2)\cos(\log(2)\theta) + \frac{2}{\sqrt{3}}\log(3)\cos(\log(3)\theta),$$

which we plot in Figure 32.2.

We will be interested in the values of θ that correspond to higher and higher *peaks* of our trigonometric series $\hat\Phi_{\leq C}(\theta)$ as $C \to \infty$. For example, the value of θ that provides the first peak of $\hat\Phi'_{\leq 3}(\theta)$ such that $|\hat\Phi'_{\leq 3}(\theta)| > 2$ is

$$\theta = 14.135375354\ldots.$$

So in Figure 32.2 we begin this exploration by plotting $\hat\Phi_{\leq 3}(\theta)$, together with its derivative, highlighting the zeroes of the derivative.

---

[1] The trigonometric series in the text – whose spectral values are the logarithms of prime powers – may also be written as

$$\sum_{m=2}^{\infty} \Lambda(m)m^{-s} + \sum_{m=2}^{\infty} \Lambda(m)m^{-\bar s}$$

for $s = \frac{1}{2} + i\theta$, where $\Lambda(m)$ is the von-Mangoldt function.

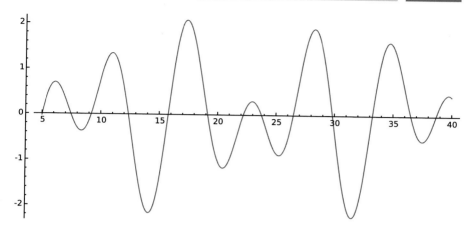

Figure 32.2. Plot of $\hat{\Phi}_{\leq 3}(\theta)$

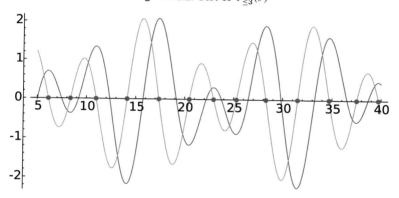

Figure 32.3. Plot of $\hat{\Phi}_{\leq 3}(\theta)$ in blue and its derivative in grey

As we shall see in subsequent figures of this chapter, there seems to be an eventual convergence of the values of θ that correspond to higher and higher *peaks*, the red dots in the figure above, as C tends to ∞. These "limit" θ-values we now insert as the endpoints of red vertical lines into Figure 32.4 comparing them with the red dots for our humble cutoff $C = 3$.

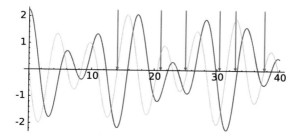

Figure 32.4. $\hat{\Phi}_{\leq 3}(\theta)$

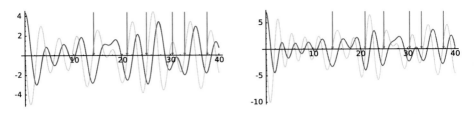

Figure 32.5. Plot of $\hat{\Phi}_{\leq C}(\theta)$ for $C = 5$ and 10

We give a further sample of graphs for a few higher cutoff values C (introducing a few more primes into the game!).

Figures 32.5–32.7 contain graphs of various cutoffs of $\hat{\Phi}_{\leq C}(\theta)$. As C increases a sequence of spikes down emerge which we indicate with red vertical arrows.

Given the numerical-experimental approach we have been adopting in this book, it is a particularly fortunate (and to us: surprising) thing that the convergence to those vertical red lines can already be illustrated using *such small* cutoff values C. One might almost imagine making hand computations that exhibit this phenomenon! Following in this spirit, see David Mumford's blog post `http://www.dam.brown.edu/people/mumford/blog/2014/RiemannZeta.html`.

To continue:

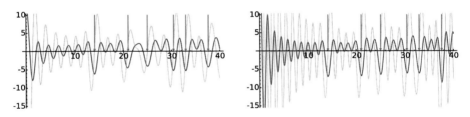

Figure 32.6. Plot of $\hat{\Phi}_{\leq C}(\theta)$ for $C = 10$ and 100

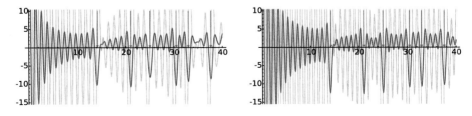

Figure 32.7. Plot of $\hat{\Phi}_{\leq C}(\theta)$ for $C = 200$ and 500

For a theoretical discussion of these spikes, see endnote [19].

The θ-coordinates of these spikes *seem to be* vaguely clustered about a discrete set of positive real numbers. These "spikes" are our first glimpse of a certain infinite set of positive real numbers

$$\theta_1, \theta_2, \theta_3, \ldots$$

which constitute the **Riemann spectrum of the primes**. If the Riemann Hypothesis holds, these numbers would be the key to the placement of primes on the number line.

By tabulating these peaks we compute – at least very approximately – ...

$$\theta_1 = 14.134725\ldots$$

$$\theta_2 = 21.022039\ldots$$

$$\theta_3 = 25.010857\ldots$$

$$\theta_4 = 30.424876\ldots$$

$$\theta_5 = 32.935061\ldots$$

$$\theta_6 = 37.586178\ldots$$

Riemann defined this sequence of numbers in his 1859 article in a manner somewhat different from the treatment we have given. In that article these θ_i appear as "imaginary parts of the nontrivial zeroes of his zeta function;" we will discuss this briefly in Part IV, Chapter 37 below.

33 How Many θ_i's are There?

The Riemann spectrum, $\theta_1, \theta_2, \theta_3, \ldots$ clearly deserves to be well understood! What do we know about this sequence of positive real numbers?

Just as we did with the prime numbers before, we can count these numbers $\theta_1 = 14.1347\ldots$, $\theta_2 = 21.0220\ldots$, etc., and form the staircase of Figure 33.1, with a step up at θ_1, a step up at θ_2, etc.

Again, just as with the staircase of primes, we might hope that as we plot this staircase from a distance as in Figures 33.2 and 33.3 that it will look like a beautiful smooth curve.

In fact, we know, conditional on RH, the staircase of real numbers $\theta_1, \theta_2, \theta_3, \ldots$ is very closely approximated by the curve

$$\frac{T}{2\pi} \log \frac{T}{2\pi e},$$

(the error term being bounded by a constant times $\log T$).

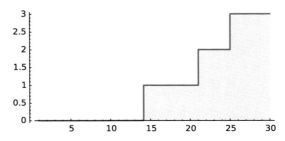

Figure 33.1. The staircase of the Riemann spectrum

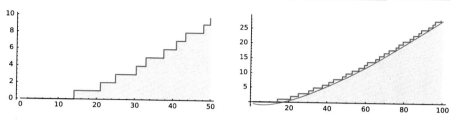

Figure 33.2. The staircase of the Riemann spectrum and the curve $\frac{T}{2\pi} \log \frac{T}{2\pi e}$

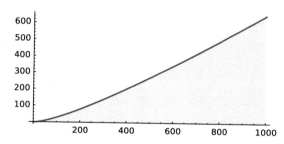

Figure 33.3. The staircase of the Riemann spectrum looks like a smooth curve

Nowadays, these mysterious numbers θ_i, these spectral lines for the staircase of primes, are known in great abundance and to great accuracy. Here is the smallest one, θ_1, given with over 1,000 digits of its decimal expansion:

14.134725141734693790457251983562470270784257115699243175685567460149963429809256764949010393171561012779202971548797436766142691469882254582505363239447137780413381237205970549621955865860200555566725836010773700205410982661507542780517442591306254481978651072304938725629738321577420395215725674809332140034990468034346267314420920377385487141378317356396995365428113079680531491688529067820822980492643386667346233200787587617920056048680543568014444246510655975686665903228686510544859444320624072727030294274522213048748720924123851418351460542790152447833835425453344004887936806761697300819000731393854983736215013045167266838920039176285123212854220523969133425832275335164060169763527563758969537674920336127209259991730427075683087951184453489180086300826483125169112710682910523759617977431815170713545316775495153828937849036474709727019948485532209253574357909226125247736595518016975233461213977316005354125926747455725877801472609830808978600712532087509395997966660675378381214891908864977277554420656532052405

and if, by any chance, you wish to peruse the first 2,001,052 of these θ_i's calculated to an accuracy within $3 \cdot 10^{-9}$, consult Andrew Odlyzko's tables:

```
http://www.dtc.umn.edu/~odlyzko/zeta_tables
```

34 Further Questions About the Riemann Spectrum

Since people have already computed[1] the first 10 trillion θ's and have never found one with multiplicity > 1, it is generally expected that the multiplicity of all the θ's in the Riemann spectrum is 1.

But, independent of that expectation, our convention in what follows will be that we *count* each of the elements in the Riemann spectrum repeated as many times as their multiplicity. So, if it so happens that θ_n occurs with multiplicity two, we view the Riemann spectrum as being the series of numbers

$$\theta_1, \theta_2, \ldots, \theta_{n-1}, \theta_n, \theta_n, \theta_{n+1}, \ldots$$

It has been conjectured that there are no infinite arithmetic progressions among these numbers. More broadly, one might expect that there is no visible correlation between the θ_i's and translation, i.e., that the distribution of θ_i's modulo any positive number T is random, as in Figure 34.1.

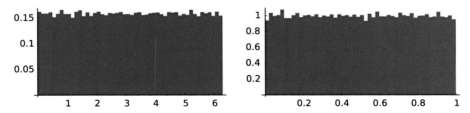

Figure 34.1. Frequency histogram of Odlyzko's computation of the Riemann spectrum modulo 2π (left) and modulo 1 (right)

[1] See http://numbers.computation.free.fr/Constants/constants.html for details.

In analogy with the discussion of prime gaps in Chapter 6, we might compute *pair correlation functions* and the statistics of gaps between successive θ's in the Riemann spectrum (see http://www.baruch.cuny.edu/math/Riemann_Hypothesis/zeta.zero.spacing.pdf). This study was begun by H. L. Montgomery and F. J. Dyson. As Dyson noticed, the distributions one gets from the Riemann spectrum bears a similarity to the distributions of eigenvalues of a random unitary matrix.[2] This has given rise to what is know as the *random matrix heuristics*, a powerful source of conjectures for number theory and other branches of mathematics.

Here is a histogram of the distribution of differences $\theta_{i+1} - \theta_i$:

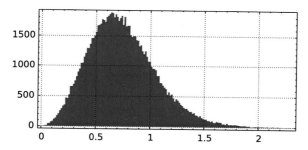

Figure 34.2. Frequency histogram of the first 99,999 gaps in the Riemann spectrum

---

[2] Any connection of the Riemann spectrum to eigenvalues of matrices would indeed be exciting for our understanding of the Riemann Hypothesis, in view of what is known as the *Hilbert-Pólya Conjecture* (see http://en.wikipedia.org/wiki/Hilbert%E2%80%93P%C3%B3lya_conjecture).

35 Going from the Riemann Spectrum to the Primes

To justify the name *Riemann spectrum of primes* we will investigate graphically whether in an analogous manner we can use this spectrum to get information about the placement of prime numbers. We might ask, for example, if there is a trigonometric function with frequencies given by this collection of real numbers,

$$\theta_1, \theta_2, \theta_3, \ldots$$

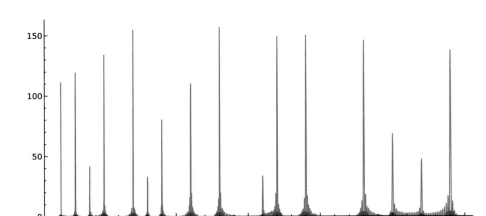

Figure 35.1. Illustration of $-\sum_{i=1}^{1000} \cos(\log(s)\theta_i)$, where $\theta_1 \sim 14.13, \ldots$ are the first 1000 contributions to the Riemann spectrum. The red dots are at the prime powers p^n, whose size is proportional to $\log(p)$

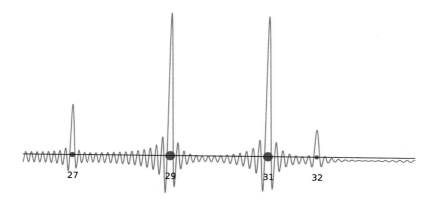

Figure 35.2. Illustration of $-\sum_{i=1}^{1000} \cos(\log(s)\theta_i)$ in the neighborhood of a twin prime. Notice how the two primes 29 and 31 are separated out by the Fourier series, and how the prime powers 3^3 and 2^5 also appear

that somehow pinpoints the prime powers, just as our functions

$$\hat{\Phi}(\theta)_{\leq C} = 2 \sum_{\text{prime powers } p^n \leq C} p^{-m/2} \log(p) \cos(n \log(p)\theta)$$

for large C pinpoint the spectrum (as discussed in the previous two chapters).

To start the return game, consider this sequence of trigonometric functions that have (*zero* and) the θ_i as spectrum

$$G_C(x) := 1 + \sum_{i<C} \cos(\theta_i \cdot x).$$

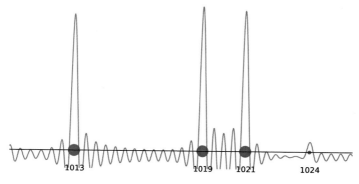

Figure 35.3. Fourier series from 1,000 to 1,030 using 15,000 of the numbers θ_i. Note the twin primes 1,019 and 1,021 and that $1,024 = 2^{10}$

As we'll see presently it is best to view these functions on a logarithmic scale so we will make the substitution of variables $x = \log(s)$ and write

$$H_C(s) := G_C(\log(s)) = 1 + \sum_{i < C} \cos(\theta_i \cdot \log(s)).$$

The theoretical story behind the phenomena that we will see graphically in this chapter is a manifestation of Riemann's explicit formula. For modern text references that discuss this general subject, see endnote [20].

PART IV

Back to Riemann

36 How to Build $\pi(X)$ from the Spectrum (Riemann's Way)

We have been dealing in Part III of our book with $\Phi(t)$ a distribution that – we said – contains all the essential information about the placement of primes among numbers. We have given a clean restatement of Riemann's hypothesis, the third restatement so far, in term of this $\Phi(t)$. But $\Phi(t)$ was the effect of a series of recalibrations and reconfigurings of the original untampered-with staircase of primes. A test of whether we have strayed from our original problem – to understand this staircase – would be whether we can return to the original staircase, and "reconstruct it" so to speak, solely from the information of $\Phi(t)$ – or equivalently, assuming the Riemann Hypothesis as formulated in Chapter 19 – can we construct the staircase of primes $\pi(X)$ solely from knowledge of the sequence of real numbers $\theta_1, \theta_2, \theta_3, \ldots$?

The answer to this is yes (given the Riemann Hypothesis), and is discussed very beautifully by Bernhard Riemann himself in his famous 1859 article.

Bernhard Riemann used the spectrum of the prime numbers to provide an exact analytic formula that analyzes and/or synthesizes the staircase of primes. This formula is motivated by Fourier's analysis of functions as constituted out of cosines. Recall from Chapter 13 that Gauss's guess is $\mathrm{Li}(X) = \int_2^X dt/\log(t)$. To continue this discussion, we do need some familiarity with complex numbers, for the definition of Riemann's exact formula requires extending the definition of the function $\mathrm{Li}(X)$ to make sense for complex numbers $X = a + bi$. In fact, more naturally, one might work with the path integral $\mathrm{li}(X) := \int_0^X dt/\log(t)$.

Riemann begins his discussion (see Figure 36.1) by defining

$$R(X) = \sum_{n=1}^{\infty} \frac{\mu(n)}{n} \mathrm{li}(X^{\frac{1}{n}}) = \lim_{N \to \infty} R^{(N)}(X) := \lim_{N \to \infty} \sum_{n=1}^{N} \frac{\mu(n)}{n} \mathrm{li}(X^{\frac{1}{n}}),$$

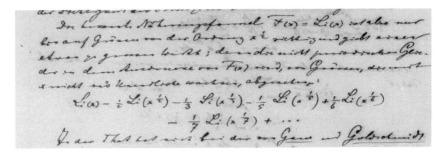

Figure 36.1. The definition of $R(X)$. Courtesy of SUB Göttingen

where $R^{(N)}(X)$ denotes the truncated sum, which one can compute as an approximation.

In all the discussion of this section the order of summation is important. For such considerations and issues regarding actual computation we refer to Riesel-Gohl (see http://wstein.org/rh/rg.pdf).

Here $\mu(n)$ is the Möbius function which is defined by

$$\mu(n) = \begin{cases} 1 & \text{if } n \text{ is a square-free positive integer with an even number of distinct prime factors,} \\ -1 & \text{if } n \text{ is a square-free positive integer with an odd number of distinct prime factors,} \\ 0 & \text{if } n \text{ is not square-free.} \end{cases}$$

See Figure 36.2 for a plot of the Möbius function.

In Chapter 17 we encountered the Prime Number Theorem, which asserts that $X/\log(X)$ and $\mathrm{Li}(X)$ are both approximations for $\pi(X)$, in the sense that both go to infinity at the same rate. That is, the ratio of any two of these three functions tends to 1 as X goes to ∞. Our first formulation of the Riemann Hypothesis (see page 41) was that $\mathrm{Li}(X)$ is an essentially square root accurate approximation of $\pi(X)$. Figures 36.3–36.4 illustrate that Riemann's function $R(X)$ appears to be an even better approximation to $\pi(X)$ than anything we have seen before.

Think of Riemann's smooth curve $R(X)$ as the *fundamental* approximation to $\pi(X)$. Riemann offered much more than just a (conjecturally) better approximation to $\pi(X)$ in his wonderful 1859 article (see Figure 36.5). He found a way

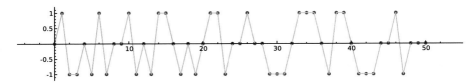

Figure 36.2. The blue dots plot the values of the Möbius function $\mu(n)$, which is only defined at integers

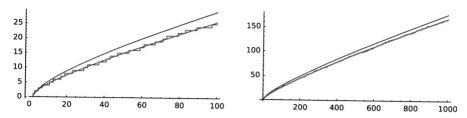

Figure 36.3. Comparisons of Li(X) (top), $\pi(X)$ (middle), and $R(X)$ (bottom, computed using 100 terms)

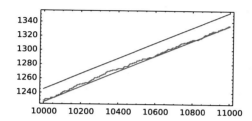

Figure 36.4. Closeup comparison of Li(X) (top), $\pi(X)$ (middle), and $R(X)$ (bottom, computed using 100 terms)

to construct what looks vaguely like a Fourier series, but with $\sin(X)$ replaced by $R(X)$ and its spectrum the θ_i, which conjecturally equals $\pi(X)$ (with a slight correction if the number X is itself a prime).

In this manner, Riemann gave an infinite sequence of improved guesses, beginning with $R_0(X)$ (see equation (18) of Riesel-Gohl at `http://wstein.org/rh/rg.pdf`) a modification of $R(X)$ that takes account of the pole and the trivial zeroes of the Riemann zeta-function, and then considered a sequence:

$$R_0(X), \quad R_1(X), \quad R_2(X), \quad R_3(X), \quad \ldots$$

and he hypothesized that one and all of them were all essentially square root approximations to $\pi(X)$, and that the sequence of these better and better approximations converge to give an exact formula for $\pi(X)$.

Thus not only did Riemann provide a "fundamental" (that is, a smooth curve that is astoundingly close to $\pi(X)$) but he viewed this as just a starting point, for

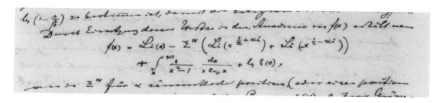

Figure 36.5. Riemann's analytic formula for $\pi(X)$. Courtesy of SUB Göttingen

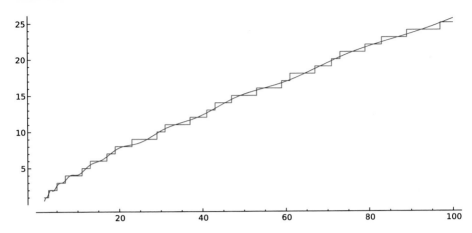

Figure 36.6. The function R_1 approximating the staircase of primes up to 100

he gave the recipe for providing an infinite sequence of corrective terms – call them Riemann's *harmonics*; we will denote the first of these "harmonics" $C_1(X)$, the second $C_2(X)$, etc. Riemann gets his first corrected curve, $R_1(X)$, from $R_0(X)$ by adding this first harmonic to the fundamental,

$$R_1(X) = R_0(X) + C_1(X),$$

he gets the second by correcting $R_1(X)$ by adding the second harmonic

$$R_2(X) = R_1(X) + C_2(X),$$

and so on

$$R_3(X) = R_2(X) + C_3(X),$$

and in the limit provides us with an exact fit.

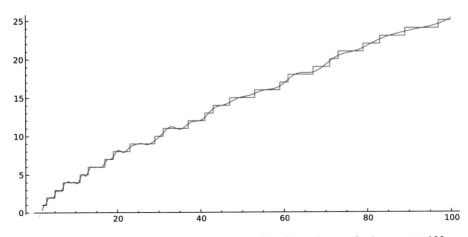

Figure 36.7. The function R_{10} approximating the staircase of primes up to 100

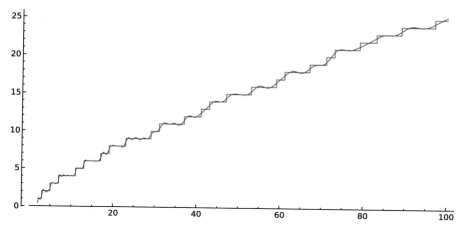

Figure 36.8. The function R_{25} approximating the staircase of primes up to 100

The Riemann Hypothesis, if true, would tell us that these correction terms $C_1(X)$, $C_2(X)$, $C_3(X)$, ... are all *square-root small.*

The elegance of Riemann's treatment of this problem is that the corrective terms $C_k(X)$ are all *modeled on* the fundamental $R(X)$ and are completely described if you know the sequence of real numbers $\theta_1, \theta_2, \theta_3, \ldots$ of the last section.

Assuming the Riemann Hypothesis, the Riemann correction terms $C_k(X)$ are defined to be

$$C_k(X) = -R(X^{\frac{1}{2}+i\theta_k}) - R(X^{\frac{1}{2}-i\theta_k}),$$

where $\theta_1 = 14.134725\ldots, \theta_2 = 21.022039\ldots$, etc., is the spectrum of the prime numbers [21].

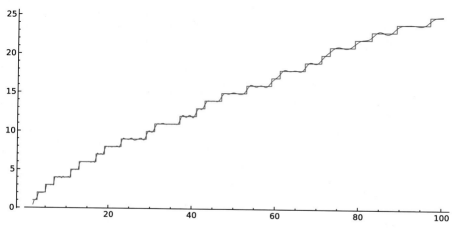

Figure 36.9. The function R_{50} approximating the staircase of primes up to 100

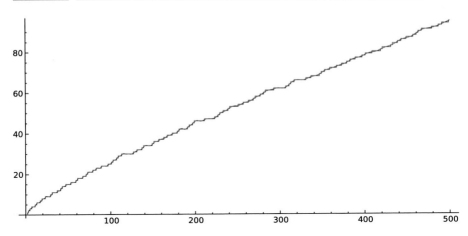

Figure 36.10. The function R_{50} approximating the staircase of primes up to 500

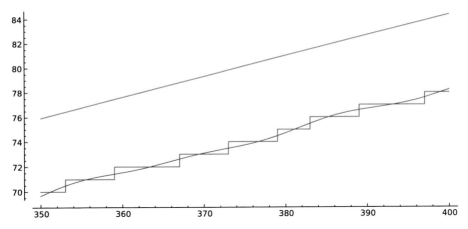

Figure 36.11. The function Li(X) (top, green), the function $R_{50}(X)$ (in blue), and the staircase of primes on the interval from 350 to 400

In sum, Riemann provided an extraordinary recipe that allows us to work out the harmonics,

$$C_1(X), \quad C_2(X), \quad C_3(X), \quad \cdots$$

without our having to consult, or compute with, the actual staircase of primes. As with Fourier's modus operandi where both *fundamental* and all *harmonics* are modeled on the sine wave, but appropriately calibrated, Riemann fashioned his higher harmonics, modeling them all on a single function, namely $R(X)$.

The convergence of $R_k(X)$ to $\pi(X)$ is strikingly illustrated in the plots in Figures 36.6–36.11 of R_k for various values of k.

As Riemann Envisioned It, the Zeta Function Relates the Staircase of Primes to Its Riemann Spectrum

37

In the previous chapters we have described – using Riemann's Hypothesis – how to obtain the *spectrum*

$$\theta_1, \theta_2, \theta_3, \ldots$$

from the staircase of primes, and hinted at how to go back. Roughly speaking, we were performing "Fourier transformations" to make this transit. But Riemann, on the very first page of his 1859 memoir, construes the relationship we have been discussing, between spectrum and staircase, in an even more profound way.

To talk about this extraordinary insight of Riemann, one would need to do two things that might seem remote from our topic, given our discussion so far.

- We will discuss a key idea that Leonhard Euler had (circa 1740).
- To follow the evolution of this idea in the hands of Riemann, we would have to assume familiarity with basic complex analysis.

We will say only a few words here about this, in hopes of giving at least a shred of a hint of how marvelous Riemann's idea is. We will be drawing, at this point, on some further mathematical background. For readers who wish to pursue the themes we discuss, here is a list of sources, that are our favorites among those meant to be read by a somewhat broader audience than people very advanced in the subject. We list them in order of "required background."

1. John Derbyshire's *Prime Obsession: Bernhard Riemann and the Greatest Unsolved Problem in Mathematics* (2003). We have already mentioned this book in our foreword, but feel that it is so good, that it is worth a second mention here.

2. The Wikipedia entry for Riemann's Zeta Function (`http://en.wikipedia.org/wiki/Riemann_zeta_function`). It is difficult to summarize who wrote this, but we feel that it is a gift to the community in its clarity. Thanks authors!

3. Enrico Bombieri's article [22]. To comprehend all ten pages of this excellent and fairly thorough account may require significant background, but try your hand at it; no matter where you stop, you will have seen good things in what you have read.

Leonhard Euler's idea ($\simeq$1740): As readers of Jacob Bernoulli's *Ars Conjectandi* (or of the works of John Wallis) know, there was in the early 18th century already a rich mathematical theory of finite sums of (non-negative k-th powers) of consecutive integers. This sum,

$$S_k(N) = 1^k + 2^k + 3^k + \cdots + (N-1)^k,$$

is a polynomial in N of degree $k+1$ with no constant term, a leading term equal to $\frac{1}{k+1}N^{k+1}$, and a famous linear term. The coefficient of the linear term of the polynomial $S_k(N)$ is the *Bernoulli number B_k*:

$$S_1(N) = 1 + 2 + 3 + \cdots + (N-1) = \frac{N(N-1)}{2} = \frac{N^2}{2} - \frac{1}{2} \cdot N,$$

$$S_2(N) = 1^2 + 2^2 + 3^2 + \cdots + (N-1)^2 = \frac{N^3}{3} + \cdots - \frac{1}{6} \cdot N,$$

$$S_3(N) = 1^3 + 2^3 + 3^3 + \cdots + (N-1)^3 = \frac{N^4}{4} + \cdots - 0 \cdot N,$$

$$S_4(N) = 1^4 + 2^4 + 3^4 + \cdots + (N-1)^4 = \frac{N^5}{5} + \cdots - \frac{1}{30} \cdot N,$$

etc. For odd integers $k > 1$ this linear term vanishes. For even integers $2k$ the Bernoulli number B_{2k} is the rational number given by the coefficient of $\frac{x^{2k}}{(2k)!}$ in the power series expansion

$$\frac{x}{e^x - 1} = 1 - \frac{x}{2} + \sum_{k=1}^{\infty} (-1)^{k+1} B_{2k} \frac{x^{2k}}{(2k)!}.$$

So

$$B_2 = \frac{1}{6}, \qquad B_4 = \frac{1}{30}, \qquad B_6 = \frac{1}{42}, \qquad B_8 = \frac{1}{30},$$

and to convince you that the numerator of these numbers is not always 1, here are a few more:

$$B_{10} = \frac{5}{66}, \qquad B_{12} = \frac{691}{2730}, \qquad B_{14} = \frac{7}{6}.$$

If you turn attention to sums of negative k-th powers of consecutive integers, then when $k = -1$,

$$S_{-1}(N) = \frac{1}{1} + \frac{1}{2} + \frac{1}{3} + \cdots + \frac{1}{N}$$

tends to infinity like log(N), but for $k < -1$ we are facing the sum of reciprocals of powers (of exponent > 1) of consecutive whole numbers, and $S_k(N)$ converges. This is the first appearance of the zeta function $\zeta(s)$ for arguments $s = 2, 3, 4, \ldots$ So let us denote these limits by notation that has been standard, after Riemann:

$$\zeta(s) := \frac{1}{1^s} + \frac{1}{2^s} + \frac{1}{3^s} + \cdots$$

The striking reformulation that Euler discovered was the expression of this infinite sum as an infinite product of factors associated to the prime numbers:

$$\zeta(s) = \sum_n \frac{1}{n^s} = \prod_{p \text{ prime}} \frac{1}{1 - p^{-s}},$$

where the infinite sum on the left and the infinite product on the right both converge (and are equal) if $s > 1$. He also evaluated these sums at even positive integers, where – surprise – the Bernoulli numbers come in again; and they and π combine to yield the values of the zeta function at even positive integers:

$$\zeta(2) = \frac{1}{1^2} + \frac{1}{2^2} + \cdots = \pi^2/6 \simeq 1.65\ldots$$

$$\zeta(4) = \frac{1}{1^4} + \frac{1}{2^4} + \cdots = \pi^4/90 \simeq 1.0823\ldots$$

and, in general,

$$\zeta(2n) = \frac{1}{1^{2n}} + \frac{1}{2^{2n}} + \cdots = (-1)^{n+1} B_{2n} \pi^{2n} \cdot \frac{2^{2n-1}}{(2n)!}.$$

A side note to Euler's formulas comes from the fact (only known much later) that no power of π is rational: do you see how to use this to give a proof that there are infinitely many primes, combining Euler's infinite product expansion displayed above with the formula for $\zeta(2)$, or with the formula for $\zeta(4)$, or, in fact, for the formulas for $\zeta(2n)$ for *any* n you choose?

Pafnuty Lvovich Chebyshev's idea ($\simeq$1845): The second moment in the history of evolution of this function $\zeta(s)$ is when Chebyshev used the *same formula* as above in the extended range where s is allowed now to be a real variable – not just an integer – greater than 1. Making use of this extension of the range of definition of Euler's sum of reciprocals of powers of consecutive whole numbers, Chebyshev could prove that for large x the ratio of $\pi(x)$ and $x/\log(x)$ is bounded above and below by two explicitly given constants. He also proved that there exists a prime number in the interval bounded by n and $2n$ for any positive integer n (this was called *Bertrand's postulate*; see http://en.wikipedia.org/wiki/Proof_of_Bertrand%27s_postulate).

Riemann's idea (1859): It is in the third step of the evolution of $\zeta(s)$ that something quite surprising happens. Riemann extended the range of Chebyshev's sum of reciprocals of positive real powers of consecutive whole numbers allowing the argument s to range over the entire complex plane s

(avoiding $s = 1$). Now this is a more mysterious extension of Euler's function, and it is deeper in two ways:

- The formula

$$\zeta(s) := \frac{1}{1^s} + \frac{1}{2^s} + \frac{1}{3^s} + \cdots$$

  does converge when the real part of the exponent s is greater than 1 (i.e., this allows us to use the same formula, as Chebyshev had done, for the right half plane in the complex plane determined by the condition $s = x + iy$ with $x > 1$ but not beyond this). You can't simply use the same formula for the extension.
- So you must face the fact that if you wish to "extend" a function beyond the natural range in which its defining formula makes sense, there may be many ways of doing it.

To appreciate the second point, the theory of a complex variable is essential. The *uniqueness* (but not yet the *existence*) of Riemann's extension of $\zeta(s)$ to the entire complex plane is guaranteed by the phenomenon referred to as *analytic continuation*. If you are given a function on any infinite subset X of the complex plane that contains a limit point, and if you are looking for a function on the entire complex plane[1] that is differentiable in the sense of complex analysis, there may be no functions at all that have that property, but if there is one, that function is *unique*. But Riemann succeeded: he was indeed able to extend Euler's function to the entire complex plane except for the point $s = 1$, thereby defining what we now call *Riemann's zeta function*.

Those ubiquitous Bernoulli numbers reappear yet again as values of this *extended* zeta function at negative integers:

$$\zeta(-n) = -B_{n+1}/(n+1)$$

so since the Bernoulli numbers indexed by odd integers > 1 all vanish, the extended zeta function $\zeta(s)$ actually vanishes at *all* even negative integers.

The even integers $-2, -4, -6, \ldots$ are often called the **trivial zeroes** of the Riemann zeta function. There are indeed other zeroes of the zeta function, and those other zeroes could – in no way – be dubbed "trivial," as we shall shortly see.

It is time to consider these facts:

1. **Riemann's zeta function codes the placement of prime powers among all numbers.** The key here is to take the logarithm and then the derivative of $\zeta(s)$ (this boils down to forming $\frac{d\zeta}{ds}(s)/\zeta(s)$). Assuming that the real part of s is > 1, taking the logarithm of $\zeta(s)$ – using Euler's infinite product formulation – gives us

$$\log\zeta(s) = \sum_{p \text{ prime}} -\log(1 - p^{-s}),$$

---

[1] or to any connected open subset that contains X.

and we can do this term-by-term, since the real part of s is > 1. Then taking the derivative gives us:

$$\frac{d\zeta}{ds}(s)/\zeta(s) = -\sum_{n=1}^{\infty} \Lambda(n)n^{-s}$$

where

$$\Lambda(n) := \begin{cases} \log(p) & \text{when } n = p^k \text{ for } p \text{ a prime number and } k > 0, \text{ and} \\ 0 & \text{if } n \text{ is not a power of a prime number.} \end{cases}$$

In particular, $\Lambda(n)$ "records" the placement of prime powers.

2. **You know lots about an analytic function if you know its zeroes and poles.**
   For example for polynomials, or even rational functions: if someone told you that a certain rational function $f(s)$ vanishes to order 1 at 0 and at ∞; and that it has a double pole at $s = 2$ and at all other points has finite nonzero values, then you can immediate say that this mystery function is a nonzero constant times $s/(s-2)^2$.

   Knowing the zeroes and poles (in the complex plane) alone of the Riemann zeta function doesn't entirely pin it down – you have to know more about its behavior at infinity since – for example, multiplying a function by e^z doesn't change the structure of its zeroes and poles in the finite plane. But a complete understanding of the zeroes and poles of $\zeta(s)$ will give all the information you need to pin down the placement of primes among all numbers.

   So here is the score:
   - As for poles, $\zeta(s)$ has only one pole. It is at $s = 1$ and is of order 1 (a "simple pole").
   - As for zeroes, we have already mentioned the trivial zeroes (at negative even integers), but $\zeta(s)$ also has infinitely many *nontrivial* zeroes. These nontrivial zeroes are known to lie in the vertical strip

   $$0 < \text{real part of } s < 1.$$

And here is yet another equivalent statement of Riemann's Hypothesis – this being the formulation closest to the one given in his 1859 memoir:

The Riemann Hypothesis (fourth formulation)

All the nontrivial zeroes of $\zeta(s)$ lie on the vertical line in the complex plane consisting of the complex numbers with real part equal to 1/2. These zeroes are none other than $\frac{1}{2} \pm i\theta_1, \frac{1}{2} \pm i\theta_2, \frac{1}{2} \pm i\theta_3, \ldots$, where $\theta_1, \theta_2, \theta_3, \ldots$ comprise the spectrum of primes we talked about in the earlier chapters.

The "$\frac{1}{2}$" that appears in this formula is directly related to the fact – correspondingly conditional on RH – that $\pi(X)$ is "square-root accurately" approximated by Li(X). That is, the error term is bounded by $X^{\frac{1}{2}+\epsilon}$. It has been conjectured that *all* the zeroes of $\zeta(s)$ are simple zeroes.

Here is how Riemann phrased RH:

> "One now finds indeed approximately this number of real roots within these limits, and it is very probable that all roots are real. Certainly one would wish for a stricter proof here; I have meanwhile temporarily put aside the search for this after some futile attempts, as it appears unnecessary for the next objective of my investigation."

In the above quotation, Riemann's roots are the θ_i's and the statement that they are "real" is equivalent to RH.

The zeta function, then, is the vise, that so elegantly clamps together information about the placement of primes and their spectrum!

That a simple geometric property of these zeroes (lying on a line!) is directly equivalent to such profound (and more difficult to express) regularities among prime numbers suggests that these zeroes and the parade of Riemann's corrections governed by them – when we truly comprehend their message – may have lots more to teach us, may eventually allow us a more powerful understanding of arithmetic. This infinite collection of complex numbers, i.e., the nontrivial zeroes of the Riemann zeta function, plays a role with respect to $\pi(X)$ rather like the role the *spectrum* of the Hydrogen atom plays in Fourier's theory. Are the primes themselves no more than an epiphenomenon, behind which there lies, still veiled from us, a yet-to-be-discovered, yet-to-be-hypothesized, profound conceptual key to their perplexing orneriness? Are the many innocently posed, yet unanswered, phenomenological questions about numbers – such as in the ones listed earlier – waiting for our discovery of this deeper level of arithmetic? Or for layers deeper still? Are we, in fact, just at the beginning?

These are not completely idle thoughts, for a tantalizing analogy relates the number theory we have been discussing to an already established branch of mathematics – due, largely, to the work of Alexander Grothendieck, and Pierre Deligne – where the corresponding analogue of Riemann's hypothesis has indeed been proved . . . .

38 Companions to the Zeta Function

Our book, so far, has been exclusively about Riemann's $\zeta(s)$ and its zeroes. We have been discussing how (the placement of) the zeroes of $\zeta(s)$ in the complex plane contains the information needed to understand (the placement of) the primes in the set of all whole numbers; and conversely.

It would be wrong – we think – if we don't even mention that $\zeta(s)$ fits into a broad family of similar functions that connect to other problems in number theory.

For example – instead of the ordinary integers – consider the *Gaussian integers*. This is the collection of numbers

$$\{a + bi\}$$

where $i = \sqrt{-1}$ and a, b are ordinary integers. We can add and multiply two such numbers and get another of the same form. The only "units" among the Gaussian integers (i.e., numbers whose inverse is again a Gaussian integer) are the four numbers ± 1, $\pm i$ and if we multiply any Gaussian integer $a + bi$ by any of these four units, we get the collection $\{a + bi, -a - bi, -b + ai, b - ai\}$. We measure the *size* of a Gaussian integer by the square of its distance to the origin, i.e.,

$$|a + bi|^2 = a^2 + b^2.$$

This size function is called the **norm** of the Gaussian integer $a + bi$ and can also be thought of as the product of $a + bi$ and its "conjugate" $a - bi$. Note that the norm is a nice multiplicative function on the set of Gaussian integers, in that the norm of a product of two Gaussian integers is the product of the norms of each of them.

We have a natural notion of **prime Gaussian integer**, i.e., one with $a > 0$ and $b \geq 0$ that cannot be factored as the product of two Gaussian integers of smaller size. Given that every nonzero Gaussian integer is uniquely expressible as a unit times a product of prime Gaussian integers, can you prove that if a Gaussian integer is a prime Gaussian integer, then its size must either be an ordinary prime number, or the square of an ordinary prime number?

Figure 38.1 contains a plot of the first few Gaussian primes as they display themselves amongst complex numbers:

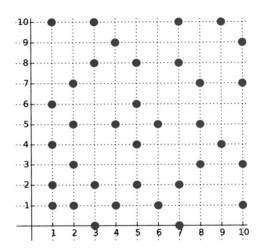

Figure 38.1. Gaussian primes with coordinates up to 10

Figure 38.2 plots a much larger number of Gaussian primes:

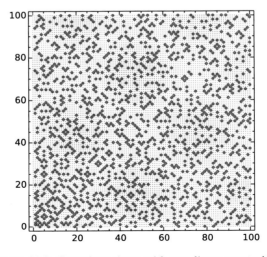

Figure 38.2. Gaussian primes with coordinates up to 100

Figures 38.3–38.6 plot the number of Gaussian primes up to each norm:

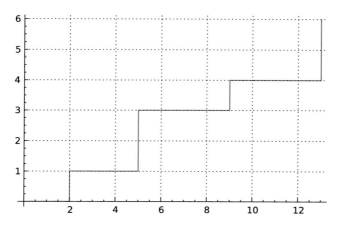

Figure 38.3. Staircase of Gaussian primes of norm up to 14

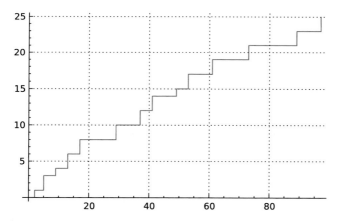

Figure 38.4. Staircase of Gaussian primes of norm up to 100

The natural question to ask, then, is: how are the Gaussian prime numbers distributed? Can one provide as close an estimate to their distribution and structure, as one has for ordinary primes? The answer, here is yes: there is a companion theory, with an analogue to the Riemann zeta function playing a role similar to the prototype $\zeta(s)$. And, it seems as if its "nontrivial zeroes" behave similarly: as far as things have been computed, they all have the property that their real part is equal to $\frac{1}{2}$. That is, we have a companion to the Riemann Hypothesis.

This is just the beginning of a much larger story related to what has been come to be called the "Grand Riemann Hypotheses" and connects to analogous problems, some of them actually solved, that give some measure

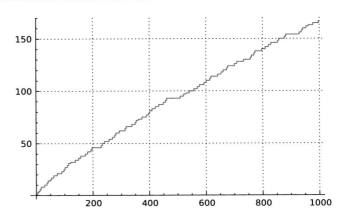

Figure 38.5. Staircase of Gaussian primes of norm up to 1000

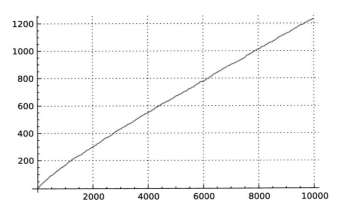

Figure 38.6. Staircase of Gaussian primes of norm up to 10000

of evidence for the truth of these hypotheses. For example, for any system of polynomials in a fixed number of variables (with integer coefficients, say) and for each prime number p there are "zeta-type" functions that contain all the information needed to count the number of simultaneous solutions in finite fields of characteristic p. That such counts can be well-approximated with a neatly small error term is related to the placement of the zeroes of these "zeta-type" functions. There is then an analogous "Riemann Hypothesis" that prescribes precise conditions on the real parts of their zeroes – this prescription being called the "Riemann Hypothesis for function fields." Now the beauty of this analogous hypothesis is that it has, in fact, been proved!

Is this yet another reason to believe the Grand Riemann Hypothesis?

Endnotes

[1] **How not to factor the numerator of a Bernoulli number:**
As mentioned in Chapter 37, the coefficient B_k of the linear term of the polynomial

$$S_k(n) = 1^k + 2^k + 3^k + \cdots + (n-1)^k$$

is (up to sign) the k-th **Bernoulli number**. These numbers are rational numbers and, putting them in lowest terms, their numerators play a role in certain problems, and their denominators in others. (This is an amazing story, which we can't go into here!)

One of us (Barry Mazur) in the recent article *How can we construct abelian Galois extensions of basic number fields?* (see http://www.ams.org/journals/bull/2011-48-02/S0273-0979-2011-01326-X/) found himself dealing (for various reasons) with the fraction $-B_{200}/400$, where B_{200} is the two-hundredth Bernoulli number. The numerator of this fraction is quite large: it is – hold your breath –

$$389 \cdot 691 \cdot 5370056528687 \qquad \text{times this 204-digit number:}$$

$N := 34526903293921580314641092817369674040684481568423967210129920642145194459192569415445652760676623601087497272415557084252765272786877636295951962087273561220060103650687168112461098659687818073890148652 7$

and he *incorrectly asserted* that it was prime. Happily, Bartosz Naskręcki spotted this error: our 204-digit N is *not* prime.

How did he know this? By using the most basic test in the repertoire of tests that we have available to check to see whether a number is prime: we'll call it the "**Fermat** 2**-test.**" We'll first give a general explanation of this type of test before we show how *N fails the Fermat 2-test.*

The starting idea behind this test is the famous result known as *Fermat's Little Theorem* where the "little" is meant to alliteratively distinguish it from you-know-what.

Theorem 39.1 (Fermat's Little Theorem)

If p is a prime number, and a is any number relatively prime to p then $a^{p-1} - 1$ is divisible by p.

A good exercise is to try to prove this, and a one-word hint that might lead you to one of the many proofs of it is *induction*.[1]

Now we are going to use this as a criterion, by – in effect – restating it in what logicians would call its *contrapositive*:

Theorem 39.2 (The Fermat a-test)

If M is a positive integer, and a is any number relatively prime to M such that $a^{M-1} - 1$ is not divisible by M, then M is not a prime.

Well, Naskręcki computed $2^{N-1} - 1$ (for the 204-digit N above) and saw that it is *not* divisible[2] by N. Ergo, our N fails the Fermat 2-test so is *not* prime.

But then, given that it is so "easy" to see that N is not prime, a natural question to ask is: what, in fact, is its prime factorization? This – it turns out – isn't so easy to determine; Naskręcki devoted 24 hours of computer time setting standard factorization algorithms on the task, and that was not sufficient time to resolve the issue. The factorization of the numerators of general Bernoulli numbers is the subject of a very interesting website run by Samuel Wagstaff (http://homes.cerias.purdue .edu/~ssw/bernoulli). Linked to this web page one finds (http:// homes.cerias.purdue.edu/~ssw/bernoulli/composite) which gives a list of composite numbers whose factorizations have resisted all attempts to date. The two-hundredth Bernoulli number is 12th on the list.

The page http://en.wikipedia.org/wiki/Integer_ factorization_records lists record challenge factorization, and one challenge that was completed in 2009 involves a difficult-to-factor number with 232 digits; its factorization was completed by a large team of researchers and took around 2000 years of CPU time. This convinced us that with sufficient motivation it would be possible to factor N, and so we asked some leaders in the field to try. They succeeded!

---

[1] Here's the proof:
$$(N + 1)^p \equiv N^p + 1 \equiv (N + 1) \mod p,$$
where the first equality is the binomial theorem and the second equality is induction.

[2] The number $2^{N-1} - 1$ has a residue after division by N of 33345811005959530251539 69739282790317394606677381970645616725285996925 66100005682927273357926209 57159782739813115005451450864072425835484898 56511276369297079926933540281 95076056916221737173183355120374 57.

Factorisation of B200
by Bill Hart on 4 Aug 05, 2012 at 07:24pm
We are happy to announce the factorization of the numerator of
the 200th Bernoulli number:

$$N = 389 \cdot 691 \cdot 5370056528687 \cdot c_{204}$$

$$c_{204} = p_{90} \cdot p_{115}$$

$$p_{90} = 149474329044343594528784250333645983079497454292$$

$$= 83824885261227075761756105767425788059260$$

$$p_{115} = 230988849487852221315416645031371036732923661613$$

$$= 61920881159759539879118404315327231419850234847$$

$$= 2629703896050377709$$

The factorization of the 204-digit composite was made possible
with the help of many people:

- William Stein and Barry Mazur challenged us to factor this
  number.
- Sam Wagstaff maintains a table of factorizations of numerators of Bernoulli numbers at http://homes.cerias.purdue
  .edu/~ssw/bernoulli/bnum. According to this table, the
  200th Bernoulli number is the 2nd smallest index with unfac-
  tored numerator (the first being the 188th Bernoulli number).
- Cyril Bouvier tried to factor the c204 by ECM up to 60-digit
  level, using the TALC cluster at Inria Nancy - Grand Est.
- yoyo@home tried to factor the c204 by ECM up to 65-digit
  level, using the help of many volunteers of the distributed
  computing platform http://www.rechenkraft.net/yoyo/.
  After ECM was unsuccessful, we decided to factor the c204 by
  GNFS.
- Many people at the Mersenne forum helped for the polynomial selection. The best polynomial was found by Shi Bai,
  using his implementation of Kleinjung's algorithm in CADO-
  NFS: http://www.mersenneforum.org/showthread.php?
  p=298264#post298264. Sieving was performed by many
  volunteers using NFS@home, thanks to Greg Childers. See
  http://escatter11.fullerton.edu/nfs for more details
  about NFS@home. This factorization showed that such a distributed effort might be feasible for a new record GNFS factorization, in particular for the polynomial selection. This was the
  largest GNFS factorization performed by NFS@home to date,
  the second largest being $2^{1040} + 1$ at 183.7 digits.
- Two independent runs of the filtering and linear algebra were
  done: one by Greg Childers with msieve (http://www.boo
  .net/~jasonp/qs.html) using a 48-core cluster made available by Bill Hart, and one by Emmanuel Thomé and Paul

Zimmermann with CADO-NFS (http://cado-nfs.gforge
.inria.fr/), using the Grid 5000 platform.
* The first linear algebra run to complete was the one with
  CADO-NFS, thus we decided to stop the other run.

Bill Hart

We verify the factorization above in SageMath as follows:

sage: p90 = 149474329044343594528784250333645983079497454292838248852612270757617561057674257880592603
sage: p115 = 23098884948785222131541664503137103673292366161361920881159759539879118404315327231419850234847626297038960503777709
sage: c204 = p90 * p115
sage: 389 * 691 * 5370056528687 * c204 == -numerator(bernoulli(200))
True
sage: is_prime(p90), is_prime(p115), is_prime(c204)
(True, True, False) .

[2] Given an integer n, there are many algorithms available for trying to write n as a product of prime numbers. First we can apply *trial division*, where we simply divide n by each prime $2, 3, 5, 7, 11, 13, \ldots$ in turn, and see what small prime factors we find (up to a few digits). After using this method to eliminate as many primes as we have patience to eliminate, we typically next turn to a technique called *Lenstra's elliptic curve method*, which allows us to check n for divisibility by bigger primes (e.g., around 10–15 digits). Once we've exhausted our patience using the elliptic curve method, we would next hit our number with something called the *quadratic sieve*, which works well for factoring numbers of the form $n = pq$, with p and q primes of roughly equal size, and n having less than 100 digits (say, though the 100 depends greatly on the implementation). All of the above algorithms – and then some – are implemented in Sage-Math, and used by default when you type factor(n) into SageMath. Try typing factor(some number, verbose=8) to see for yourself.

If the quadratic sieve fails, a final recourse is to run the *number field sieve* algorithm, possibly on a supercomputer. To give a sense of how powerful (or powerless, depending on perspective!) the number field sieve is, a record-setting factorization of a general number using this algorithm is the factorization of a 232 digit number called RSA-768 (see https://eprint.iacr.org/2010/006.pdf):

$n = 12301866845301177551304949583849627207728535695953347921973$
$22452151726400507263657518745202199786469389956474942774063845$
$92519255732630345373154826850791702612214291346167042921431160$
$2221240479274737794080665351419597459856902143413$

which factors as pq, where

$p = 33478071698956898786044169848212690817704794983713768568912$
$43138898288379387800228761471165253174308773781446799489$

and

$q = 367460436667995904282446337996279526322791581643430876426760$
$3228381573966651127923337341714339681027009279873630891 7.$

We encourage you to try to factor n in SageMath, and see that it fails. Sage-Math does not yet include an implementation of the number field sieve algorithm, though there are some free implementations currently available (see http://www.boo.net/~jasonp/qs.html).

[3] We can use SageMath (at http://sagemath.org) to quickly compute the "hefty number" $p = 2^{43,112,609} - 1$. Simply type p = 2^43112609 - 1 to instantly compute p. In what sense have we *computed p*? Internally, p is now stored in base 2 in the computer's memory; given the special form of p it is not surprising that it took little time to compute. Much more challenging is to compute all the base 10 digits of p, which takes a few seconds: d = str(p). Now type d[-50:] to see the last 50 digits of p. To compute the sum 58416637 of the digits of p type sum(p.digits()).

[4] In contrast to the situation with factorization, testing integers of this size (e.g., the primes p and q) for primality is relatively easy. There are fast algorithms that can tell whether or not any random thousand digit number is prime in a fraction of second. Try for yourself using the SageMath command is_prime. For example, if p and q are as in endnote 2, then is_prime(p) and is_prime(q) quickly output True and is_prime(p*q) outputs False. However, if you type factor(p*q, verbose=8) you can watch as SageMath tries forever and fails to factor pq.

[5] In Sage, the function prime_range enumerates primes in a range. For example, prime_range(50) outputs the primes up to 50 and prime_range(50,100) outputs the primes between 50 and 100. Typing prime_range(10^8) in SageMath enumerates the primes up to a hundred million in around a second. You can also enumerate primes up to a billion by typing v=prime_range(10^9), but this will use a large amount of memory, so be careful not to crash your computer if you try this. You can see that there are $\pi(10^9) = 50,847,534$ primes up to a billion by then typing len(v). You can also compute $\pi(10^9)$ directly, without enumerating all primes, using the command prime_pi(10^9). This is much faster since it uses some clever counting tricks to find the number of primes without actually listing them all.

In Chapter 19 we tinkered with the staircase of primes by first counting both primes and prime powers. There are comparatively few prime powers that are not prime. Up to 10^8, only 1,405 of the 5,762,860 prime powers are not themselves primes. To see this, first enter a = prime_pi(10^8); pp = len(prime_powers(10^8)). Typing (a, pp, pp-a) then outputs the triple (5761455, 5762860, 1405).

[6] Hardy and Littlewood give a nice conjectural answer to such questions about gaps between primes. See Problem **A8** of Guy's book *Unsolved Problems in Number Theory* (2004). Note that Guy's book discusses counting the number $P_k(X)$ of pairs of primes up to X that differ by a fixed even number k; we have $P_k(X) \geq \text{Gap}_k(X)$, since for $P_k(X)$ there is no requirement that the pairs of primes be consecutive.

[7] If $f(x)$ and $g(x)$ are real-valued functions of a real variable x such that for any $\epsilon > 0$ both of them take their values between $x^{1-\epsilon}$ and $x^{1+\epsilon}$ for x sufficiently large, then say that $f(x)$ and $g(x)$ are **good approximations of one another** if, for any positive ϵ the absolute value of their difference is less than $x^{\frac{1}{2}+\epsilon}$ for x sufficiently large. The functions $\mathrm{Li}(X)$ and $R(X)$ are good approximations of one another.

[8] This computation of $\pi(X)$ was done by David J. Platt in 2012, and is the largest value of $\pi(X)$ ever computed. See http://arxiv.org/abs/1203.5712 for more details.

[9] In fact, the Riemann Hypothesis is equivalent to $|\mathrm{Li}(X) - \pi(X)| \leq \sqrt{X}\log(X)$ for all $X \geq 2.01$. See Section 1.4.1 of Crandall and Pomerance book Prime Numbers: A Computational Perspective.

[10] For a proof of this here's a hint. Compute the difference between the derivatives of $\mathrm{Li}(x)$ and of $x/\log x$. The answer is $1/\log^2(x)$. So you must show that the ratio of $\int_2^X dx/\log^2(x)$ to $\mathrm{Li}(x) = \int_2^X dx/\log(x)$ tends to zero as x goes to infinity, and this is a good calculus exercise.

[11] See http://www.maths.tcd.ie/pub/HistMath/People/Riemann/Zeta/ for for the original German version and an English translation.

[12] We have

$$\psi(X) = \sum_{p^n \leq X} \log p$$

where the summation is over prime powers p^n that are $\leq X$.

[13] See http://en.wikipedia.org/wiki/Fast_Fourier_transform.

[14] http://en.wikipedia.org/wiki/Distribution_%28mathematics%29 contains a more formal definition and treatment of distributions. Here is Schwartz's explanation for his choice of the word *distribution:*

> "Why did we choose the name distribution? Because, if μ is a measure, i.e., a particular kind of distribution, it can be considered as a distribution of electric charges in the universe. Distributions give more general types of electric charges, for example dipoles and magnetic distributions. If we consider the dipole placed at the point a having magnetic moment M, we easily see that it is defined by the distribution $-D_M\delta_{(a)}$. These objects occur in physics. Deny's thesis, which he defended shortly after, introduced electric distributions of finite energy, the only ones which really occur in practice; these objects really are distributions, and do not correspond to measures. Thus, distributions have two very different aspects: they are a generalization of the notion of function, and a generalization of the notion of distribution of electric charges in space. [...] Both these interpretations of distributions are currently used."

[15] David Mumford suggested that we offer the following paragraph from http://en.wikipedia.org/wiki/Dirac_delta_function on the Dirac delta function:

> An infinitesimal formula for an infinitely tall, unit impulse delta function (infinitesimal version of Cauchy distribution) explicitly

appears in an 1827 text of Augustin Louis Cauchy. Siméon Denis
Poisson considered the issue in connection with the study of wave
propagation as did Gustav Kirchhoff somewhat later. Kirchhoff and
Hermann von Helmholtz also introduced the unit impulse as a limit
of Gaussians, which also corresponded to Lord Kelvin's notion of
a point heat source. At the end of the 19th century, Oliver Heavi-
side used formal Fourier series to manipulate the unit impulse. The
Dirac delta function as such was introduced as a "convenient notation" by Paul Dirac in his influential 1930 book *Principles of Quan-
tum Mechanics*. He called it the "delta function" since he used it as a
continuous analogue of the discrete Kronecker delta.

[16] As discussed in `http://en.wikipedia.org/wiki/Distribution_`
`%28mathematics%29`, "generalized functions" were introduced by Sergei
Sobolev in the 1930s, then later independently introduced in the late
1940s by Laurent Schwartz, who developed a comprehensive theory of
distributions.

[17] If the Riemann Hypothesis holds they are precisely the *imaginary parts of
the "nontrivial" zeroes of the Riemann zeta function.*

[18] *The construction of* $\Phi(t)$ *from* $\psi(X)$*:*
Succinctly, for positive real numbers t,

$$\Phi(t) := e^{-t/2}\Psi'(t),$$

where $\Psi(t) = \psi(e^t)$ (see Figure 39.2), and Ψ' is the derivative of $\Psi(t)$,
viewed as a distribution. We extend this function to all real arguments t
by requiring $\Phi(t)$ to be an even function of t, i.e., $\Phi(-t) = \Phi(t)$. But, to
review this at a more leisurely pace,

1. Distort the X-axis of our staircase by replacing the variable X by e^t to
get the function

$$\Psi(t) := \psi(e^t).$$

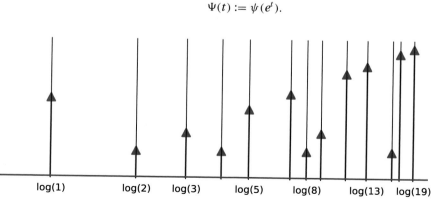

log(1)   log(2) log(3)   log(5)   log(8)   log(13) log(19)

Figure 39.1. $\Psi'(t)$ is a (weighted) sum of Dirac delta functions at the logarithms
of prime powers p^n weighted by $\log(p)$ (and by $\log(2\pi)$ at 0). The taller the arrow,
the larger the weight.

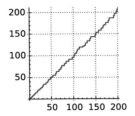

Figure 39.2. Illustration of the staircase $\psi(X)$ constructed in Chapter 19 that counts weighted prime powers.

No harm is done by this for we can retrieve our original $\psi(X)$ as

$$\psi(X) = \Psi(\log(X)).$$

Our distorted staircase has risers at (0 and) all positive integral multiples of logs of prime numbers.

2. Now we'll do something that might seem a bit more brutal: *take the derivative of this distorted staircase* $\Psi(t)$. This derivative $\Psi'(t)$ is a *generalized* function with support at all nonnegative integral multiples of logs of prime numbers.

3. Now – for normalization purposes – multiply $\Psi'(t)$ by the function $e^{-t/2}$ which has no effect whatsoever on the support.

In summary: The generalized function that resulted from the above carpentry:

$$\Phi(t) = e^{-t/2}\Psi'(t),$$

retains the information we want (the placement of primes) even if in a slightly different format.

[19] A version of the Riemann–von Mangoldt explicit formula gives some theoretical affirmation of the phenomena we are seeing here. We thank Andrew Granville for a discussion about this.

Figure 39.3. Andrew Granville. Photo by David Imms

Even though the present endnote is not the place to give anything like a full account, we can't resist setting down a few of Granville's comments that might be helpful to people who wish to go further. (This discussion can be modified to analyze what happens unconditionally, but we will be assuming the Riemann Hypothesis below.) The function $\hat{\Phi}_{\leq C}(\theta)$ that we are graphing in this chapter can be written as:

$$\hat{\Phi}_{\leq C}(\theta) = \sum_{n \leq C} \Lambda(n) n^{-w}$$

where $w = \frac{1}{2} + i\theta$. This function, in turn, may be written (by Perron's formula) as

$$\frac{1}{2\pi i} \lim_{T \to \infty} \int_{s=\sigma_0-iT}^{s=\sigma_0+iT} \sum_n \Lambda(n) n^{-w} \left(\frac{C}{n}\right)^s \frac{ds}{s}$$

$$= \frac{1}{2\pi i} \lim_{T \to \infty} \int_{s=\sigma_0-iT}^{s=\sigma_0+iT} \sum_n \Lambda(n) n^{-w-s} C^s \frac{ds}{s}$$

$$= -\frac{1}{2\pi i} \lim_{T \to \infty} \int_{s=\sigma_0-iT}^{s=\sigma_0+iT} \left(\frac{\zeta'}{\zeta}\right)(w+s) \frac{C^s}{s} ds.$$

Here, we assume that σ_0 is sufficiently large and C is not a prime power.

One proceeds, as is standard in the derivation of Explicit Formulae, by moving the line of integration to the left, picking up residues along the way. Fix the value of $w = \frac{1}{2} + i\theta$ and consider

$$K_w(s, C) := \frac{1}{2\pi i} \left(\frac{\zeta'}{\zeta}\right)(w+s) \frac{C^s}{s},$$

which has poles at

$$s = 0, \quad 1 - w, \quad \text{and} \quad \rho - w,$$

for every zero ρ of $\zeta(s)$. We distinguish five cases, giving descriptive names to each:

1. *Singular pole:* $s = 1 - w$.
2. *Trivial poles:* $s = \rho - w$ with ρ a trivial zero of $\zeta(s)$.
3. *Oscillatory poles:* $s = \rho - w = i(\gamma - \theta) \neq 0$ with $\rho = 1/2 + i\gamma (\neq w)$ a nontrivial zero of $\zeta(s)$. (Recall that we are assuming the Riemann Hypothesis, and our variable $w = \frac{1}{2} + i\theta$ runs through complex numbers of real part equal to $\frac{1}{2}$. So, in this case, s is purely imaginary.)
4. *Elementary pole:* $s = 0$ when w is not a nontrivial zero of $\zeta(s)$ – i.e., when $0 = s \neq \rho - w$ for any nontrivial zero ρ.
5. *Double pole:* $s = 0$ when w is a nontrivial zero of $\zeta(s)$ – i.e., when $0 = s = \rho - w$ for some nontrivial zero ρ. This, when it occurs, is indeed a double pole, and the residue is given by $m \cdot \log C + \epsilon$. Here m is the multiplicity of the zero ρ (which we expect always – or at least usually – to be equal to 1) and ϵ is a constant (depending on ρ, but not on C).

The standard technique for the "Explicit formula" will provide us with a formula for our function of interest $\hat{\Phi}_{\leq C}(\theta)$. The formula has terms resulting from the residues of each of the first three types of poles, and of the *Elementary* or the *Double* pole – whichever exists. Here is the shape of the formula, given with terminology that is meant to be evocative:

$$(\mathbf{1}) \quad \hat{\Phi}_{\leq C}(\theta) = \mathrm{Sing}_{\leq C}(\theta) + \mathrm{Triv}_{\leq C}(\theta) + \mathrm{Osc}_{\leq C}(\theta) + \mathrm{Elem}_{\leq C}(\theta)$$

Or:

$$(\mathbf{2}) \quad \hat{\Phi}_{\leq C}(\theta) = \mathrm{Sing}_{\leq C}(\theta) + \mathrm{Triv}_{\leq C}(\theta) + \mathrm{Osc}_{\leq C}(\theta) + \mathrm{Double}_{\leq C}(\theta),$$

the first if w is not a nontrivial zero of $\zeta(s)$ and the second if it is.

The good news is that the functions $\mathrm{Sing}_{\leq C}(\theta), \mathrm{Triv}_{\leq C}(\theta)$ (and also $\mathrm{Elem}_{\leq C}(\theta)$ when it exists) are smooth (easily describable) functions of the two variables C and θ; for us, this means that they are not that related to the essential information-laden *discontinuous* structure of $\hat{\Phi}_{\leq C}(\theta)$. Let us bunch these three contributions together and call the sum $\mathrm{Smooth}(C, \theta)$, and rewrite the above two formulae as:

$$(\mathbf{1}) \quad \hat{\Phi}_{\leq C}(\theta) = \mathrm{Smooth}(C, \theta) + \mathrm{Osc}_{\leq C}(\theta)$$

Or:

$$(\mathbf{2}) \quad \hat{\Phi}_{\leq C}(\theta) = \mathrm{Smooth}(C, \theta) + \mathrm{Osc}_{\leq C}(\theta) + m \cdot \log C + \epsilon,$$

depending upon whether or not w is a nontrivial zero of $\zeta(s)$.

We now focus our attention on the Oscillatory term, $\mathrm{Osc}_{\leq C}(\theta)$, approximating it by a truncation:

$$\mathrm{Osc}_w(C, X) := 2 \sum_{|\gamma| < X} \frac{e^{i \log C \cdot (\gamma - \theta)}}{i(\gamma - \theta)}.$$

Here if a zero has multiplicity m, then we count it m times in the sum. Also, in this formula we have relegated the "θ" to the status of a subscript (i.e., $w = \frac{1}{2} + i\theta$) since we are keeping it constant, and we view the two variables X and C as linked in that we want the cutoff "X" to be sufficiently large, say $X \gg C^2$, so that the error term can be controlled.

At this point, we can perform a "multiplicative version" of a Cesàro summation – i.e., the operator $F(c) \mapsto (\mathrm{Cés}F)(C) := \int_1^C F(c)dc/c$. This has the effect of forcing the oscillatory term to be bounded as C tends to infinity.

This implies that for any fixed θ,

- $\mathrm{Cés}\hat{\Phi}_{\leq C}(\theta)$ is bounded independent of C if θ is not the imaginary part of a nontrivial zero of $\zeta(s)$ and
- $\mathrm{Cés}\hat{\Phi}_{\leq C}(\theta)$ grows as $\frac{m}{2} \cdot (\log C)^2 + O(\log C)$ if θ is the imaginary part of a nontrivial zero of $\zeta(s)$ of multiplicity m,

giving a theoretical confirmation of the striking feature of the graphs of our Chapter 32.

[20] A reference for this is:

> [I-K]: H. Iwaniec; E. Kowalski, **Analytic Number Theory**, *American Math-ematical Society Colloquium Publications* **53** (2004).
> (See also the bibliography there.)

Many ways of seeing the explicit relationship are given in Chapter 5 of [I-K]. For example, consider Exercise 5 on page 109.

$$\sum_{\rho} \hat{\phi}(\rho) = -\sum_{n \geq 1} \Lambda(n)\phi(n) + I(\phi),$$

where

- ϕ is any smooth complex-valued function on $[1, +\infty)$ with compact support,
- $\hat{\phi}$ is its Mellin transform

$$\hat{\phi}(s) := \int_0^\infty \phi(x)x^{s-1}dx,$$

- the last term on the right, $I(\phi)$, is just

$$I(\phi) := \int_1^\infty (1 - \frac{1}{x^3 - x})\phi(x)dx$$

(coming from the pole at $s = 1$ and the "trivial zeroes").
- The more serious summation on the left hand side of the equation is over the nontrivial zeroes ρ, noting that if ρ is a nontrivial zero so is $\bar{\rho}$.

Of course, this "explicit formulation" is not *immediately* applicable to the graphs we are constructing since we cannot naively take $\hat{\phi}$ to be a function forcing the left hand side to be $G_C(x)$.

See also Exercise 7 on page 112, which discusses the sums

$$x - \sum_{|\theta| \leq C} \frac{x^{\frac{1}{2}+i\theta} - 1}{\frac{1}{2} + i\theta}.$$

[21] You may well ask how we propose to order these correction terms if RH is false. Order them in terms of (the absolute value of) their imaginary part, and in the unlikely situation that there is more than one zero with the same imaginary part, order zeroes of the same imaginary part by their real parts, going from right to left.

[22] Bombieri, *The Riemann Hypothesis*, available at http://www.claymath .org/sites/default/files/official_problem_description .pdf.

Index

Printed in the United States
By Bookmasters